Make Your House
Radon Free

Dedication

To George Dickel Tennessee Whiskey, which capped off some late nights of writing. It was always smooth, mellow, and satisfying, like finding the right words when you have been searching for them from the soul. Dickel and good books go together. Each time we celebrate the completion of a book, Dickel is the toasting drink. To Dickel—it beats the bad guy dressed in black every drink...

and to...

A Brother, the Brother Word Processor! Brother International Corp. has the ultimate word processor. It processes all words with beautiful precision, never hesitating in its performance. The screen displays pages exactly as they appear. The built-in printer produces them with superb letter quality. The Brother is personal, the personalized word processor that should be the essence of every office, at home or in town. Compact but with a big screen, a small footprint but a 64,000 character memory, portable but carries a 70,000-word dictionary. Whether you are writing a book, story, letter, invoice, or other papers, he's your BROTHER, and he's heavy on work and light to handle.

Make Your House
Radon Free

DRS. CARL and BARBARA GILES

TAB BOOKS Inc.
Blue Ridge Summit, PA

FIRST EDITION
FIRST PRINTING

Library of Congress Cataloging in Publication Data

Giles, Carl H.
    Make your house radon free / by Carl and Barbara Giles.

        p.   cm.
    ISBN 0-8306-9291-6   —ISBN 0-8306-3291-3 (pbk.)

        1. Radon—Safety measures.   2. Radon—Environmental measures.
3. Indoor air pollution.   I. Giles, Barbara, 1944-.   .   II. Title.

    TD885.5.R33G55   1989   89-36608
    628.5'35—dc20   CIP

Acquisitions Editor: Kim Tabor
Book Editor: Joanne M. Slike
Production: Katherine Brown
Paperbound Cover Design: Lori E. Schlosser

Contents

Acknowledgments   *vii*

Introduction   *viii*

1 **Lung Cancer and the Invisible Gas**   *1*

How Radon Gets into Buildings • Radon
Measurements • How Radon Kills • Government
Health Warnings • Smokers Versus Nonsmokers

2 **Testing the Radon Level in Your House**   *8*

Electret-Passive Environmental Radon Monitor •
Passive Measuring • Screening Measurements
• Follow-Up Measurements • EPA Action Level •
Charcoal Lab Analysis • Radon One Monitor •
Terradex Alpha-Track Detector Kits • Electrets
• Continuous Radon Monitors • Tamper-Resistant
Monitor Cages

3 **Radon Entry and Its Behavior**   *22*

Other Means of Radon Entry • Air Pressure
Differences • Man-Built Pathways •
Radon-Contaminated Materials • Foundations •
Indoor Radon Concentrations • Ventilation and
Radon

4   **Removing Radon by Redirection**   *35*
    Depressurization Applications • Sub-Slab Soil
    Ventilation • Sealing • Drain Tile • Soil Ventilation
    • Depressurizing Block Walls • Baseboard
    Depressurization • Depressurization under Plastic
    Films • Basement Pressurization • Sealing
    Standard Drains • Sump Pumps • Direct Vent
    Heating Systems • Apollo Coils • Matched
    Components • Direct Vent Fireplace

5   **Reducing Radon Levels by Dilution**   *71*
    Naturally Powered Ventilation • Passive Ventilation
    • Attic Ventilation • Power Ventilators • Soffit and
    Undereave Vents • Ventilation through a Warm-Air
    Furnace • Heat Recovery Ventilators

6   **Removing Radon by Filtration**   *84*
    Ionization and Radon • Thurmond Air Quality
    Systems • Ozone • Central Vacuum System •
    Molecular Adsorber • Monitors to Check Other
    Pollutants

7   **Radon and Your Water Supply**   *95*
    Private Wells • Radon in Drinking Water • Water to
    Air Radon Transfer • Measuring Radon in Water •
    Aeration Methods

8   **Building a Radon-Resistant House**   *104*
    Buyers Demand Testing • Building Sites •
    Foundation Walls • Sealing Masonry Walls • EPA
    Manual on New House Construction • Vapor-Soil
    Barriers under Concrete • Insulation and Barriers
    for Crawl Spaces • Polyurethane Sealant • AFM
    Radon Gas Abatement System Sealant

    Appendix—Compiling the Facts   *116*
    Home Inspection Survey • EPA Survey Form

    Index   *132*

Acknowledgments

EPA manuals and literature were of major assistance in providing information and line drawings for this book. Jed Harrison of the EPA Office in Washington, D.C. was most helpful in providing information and insights into radon. He was most cooperative in sharing his knowledge during telephone conversations, also.

Terry Brennan graciously contributed several excellent illustrations, and we were delighted to learn that he has also worked with wildlife rehabilitation. Allen Adams of the New York State Energy Office was most cooperative in assisting with illustrations and information.

The EPA Office in Research Triangle Park, NC was also helpful in gathering research and materials. D. Bruce Henschel, Ronald B. Mosley, and Michael C. Osborne of this office have authored some of the most informative EPA manuals. Brenda M. was also most helpful in seeing we received some materials.

Jim Schwartz of the Sika Corporation was of invaluable assistance in furnishing information. He is indeed the "king of caulking knowledge" and one of the world's authorities on concrete. Nester Noe is to be commended for his fine line of environmentally safe products, manufactured by his AFM Enterprises, Inc. His AFM Radon Gas Abatement System—sealer—is the best.

Others contributed, and they are very much thanked for their assistance.

Introduction

As this book was being completed, a Congressional House Sub-committee had just reported that 101.8 million Americans are breathing unsafe air!

The pollutants are being released in the outside air. Of course, much of this outdoor pollution comes inside homes and is magnified many times in its deadliness. And these toxic substances are *in addition* to radon! Congress passed the Clean Air Act in 1970, and many amendments have been made to it.

Six pollutants in particular have been singled out many times, and all but two have increased:

- **Particulates** include dust, soot, smoke, and particulate-laden water drops.
- **Carbon monoxide** from auto exhausts is one major contributor. It is especially deadly to the young and the elderly.
- **Sulfur dioxide** comes primarily from coal-burning plants and causes acid rain. It hits the lungs and causes respiratory tract damage.
- **Ozone** is the most pervasive air pollution problem, mostly as a result of auto and truck emissions. It irritates the eyes and is especially damaging to the young, the elderly, and the infirmed.
- **Nitrogen dioxide**, an industrial pollutant, has been cleaned up considerably across the nation. Only the Los Angeles area of California still violates federal standards.
- **Lead** has decreased in the atmosphere considerably since unleaded gasoline. In the last decade, lead levels have gone

down 88 percent. Lead is a poisonous metal, and many industrial plants still pollute the air with it. It is among the most deadly pollutants, and can cause respiratory and circulatory problems, and lung and kidney damage. It is especially deadly to children under the age of five.

Texas, Louisiana, Tennessee, Virginia, Ohio, Michigan, Illinois, and Indiana are the worst offenders in that order—each state putting more than 100 million pounds of toxic chemicals yearly into the air.

Each year, U.S. industries emit 148,181 pounds of methyl isocyanate, the chemical that killed thousands near an industrial plant in Bhopal, India in 1984. Phosgene, a World War I nerve gas, is sent into the air—69,000 pounds of it each year by Kansas industries. Texas is the biggest polluter, Hawaii the least.

Yet some public health authorities say that radon is by far the most deadly of all pollutants. Awareness of the radon threat has been steadily increasing among the public, and the U.S. Government has clearly stated its intentions and goals toward reducing radon levels.

CONGRESS TAKES ACTION ON RADON

During the Second Session of the One-Hundredth Congress of the United States, on January 25, 1988, an amendment to the Toxic Substance Control Act was passed to assist states in responding to the threat of radon exposure to human health.

"The national long-term goal of the United States with respect to radon levels in buildings is that the air within buildings in the United States should be as free of radon as the ambient air outside of buildings," the Indoor Radon Abatement states.

What can the consumer do toward achieving this common goal? Your first step is to educate yourself about radon: where it comes from, why it is dangerous, what levels are unsafe, how radon levels are measured, how to remove radon, and how to prevent its entry. Knowledge is your greatest weapon against this deadly gas.

RADON RIP-OFFS AND HOW TO AVOID THEM

Learning about radon will also help you steer clear of the all-too-frequent consumer rip-offs in radon detection and prevention. Anyone who is considering hiring someone to reduce the radon in a house should be cautious; EPA personnel have published many stories on radon rip-offs.

At one EPA radon symposium, one of the most talked about scams focused on the people who posed as radon testing experts. They would knock on a door and offer to give a five-minute radon test. After going into the basement or another area of the house, they would take an empty bottle, remove the lid and then put it back on. When they had obtained their "air sample," they would go to their van and use special equipment to get a radon reading. They would report the level was safe, collect $25 or so, and quickly leave.

Some states have already set requirements for professional testing and mitigation firms. Still, homeowners must be wary. Three estimates should be obtained only after extensive testing of different types has been done, and the problems with the structure identified as specifically as possible. These writers have seen ads in "professional" radon publications advertising information on how professional mitigation firms may increase their profits by 50 to 95 percent by using their advice!

Threshold Technical Products, Inc. is among those firms in the radon industry that is most attentive to quality products and quality control. It works to improve the ease and efficiency of continuous radon monitoring with products like the Survivor 2. Threshold president John W. Grether believes the public should have access to the best continuous radon monitoring at reasonable prices.

ABOUT THIS BOOK

Millions of homes and buildings throughout America have life-threatening radon levels. *Make Your House Radon Free* outlines in detail the techniques and mitigation methods to reduce radon entry into structures. It also specifies how to build the most radon-resistant houses and buildings. The highest-quality products that help prevent radon penetration are recommended in many instances throughout the text. The authors have amassed one of the most comprehensive information centers on radon, and our research continues. Please feel free to contact us by sending a self-addressed stamped envelope to Radon Product Research, P.O. Box 155, Black Mountain, NC 28711.

1

Lung Cancer and the Invisible Gas

God gave us radon on the second day, according to Genesis. "And God said, 'Let the waters under the heaven be gathered together unto one place, and let the dry land appear: and it was so,'" Genesis 1:9 states.

"And God saw that it was good," the next verse says. But, as the cliché notes, too much of any good thing is bad. God did not intend for modern man to let too much radon into his dwellings.

Radon gas found in the ground is a product of the natural decay of radium-226. This radioactive chemical element is just naturally present in nature and is in trace levels in most soils and many types of rocks.

Uranium decays through a chain of radioactive elements. As it deteriorates, it releases radioactive particles and electromagnetic radiation in the process. Each element or by-product in the chain is a solid except radon-222 (radon), which is a gas.

Radon is going into the air right now as it always has and always will. But the amount is so small in proportion to all outdoors, it presents no health problems.

HOW RADON GETS INTO BUILDINGS

The amount of radon gas that enters a structure depends on five factors: how much radon gas or radon parent compounds (from decaying rocks) are found in the soil beneath the house, the permeability of the soil, the presence of faults and fissures in the underlying or nearby rock, the openings between the house and soil, and the driving forces that move soil gas containing radon along these pathways into the house.

One cannot have a radon problem unless there is radium nearby, a route for the gas to move through the soil or rock, a driving force, and openings in the foundation. Since radon is always in the air, and some air is always inside a structure, some radon gas is always present.

RADON MEASUREMENTS

The U.S. Environmental Protection Agency (EPA) says that as long as there are no more than four picocuries of radon gas per liter of air, radon is not dangerous. A *picocurie* (pCi) is a unit of measurement of radioactivity. A *curie* is the amount of any radionuclide that undergoes exactly 3.7×10^{10} radioactive disintegrations per second. A picocurie is one-trillionth of a curie, or 0.037 radioactive disintegrations per second.

Picocurie per liter is a common unit of measurement of the concentration of radioactivity in a gas. A picocurie per liter corresponds to 0.037 radioactive disintegrations per second in every liter of air. To put that in a rough perspective of size and space, imagine the amount of radon gas on the tip of a toothpick inside a liter bottle, such as those that contain soft drinks.

When the radon level inside a structure is more than four picocuries per liter of air, people living or working inside may develop lung cancer. Radon is the second leading cause of lung cancer. Health authorities estimate 20,000 to 30,000 will die of lung cancer caused by radon this year. That tallies out to an average of 60 to 90 deaths daily (FIG. 1-1).

HOW RADON KILLS

Radon decays with a half-life of 3.82 days into a series of solid, short-lived radioisotopes. These are collectively called ''radon daughters'' or ''radon progeny.'' Two of the isotopes, polonium-218 and polonium-214, produce alpha particles. When people inhale these particles of radon progeny decay into the lungs in too-high levels for a period of time, the cells lining the airways may be damaged, resulting in lung cancer.

For more than 100 years, man has known that mining radioactive ores is associated with a greatly increased risk of lung cancer. Uranium miners in America, France, Canada, and Czechoslovakia have a high incidence of lung cancer.

2

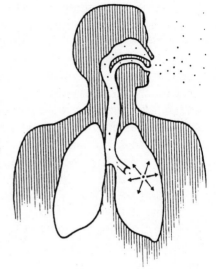

Fig. 1-1. As you breathe, radon decay products can become trapped in the lungs, and as these decay products break down further, they release small bursts of energy that can damage lung tissue and lead to lung cancer.

Other miners are also very prone to lung cancer. British and French iron miners, Chinese and British tin miners, American and Swedish metal miners, and other miners have a high incidence of lung cancer. Numerous studies also document that laboratory animals develop lung cancer after certain exposures and doses of radon gas.

GOVERNMENT HEALTH WARNINGS

In September 1988, the U.S. Federal Government issued a national health advisory that stated, "Millions of homes have elevated radon levels." It called on people nationwide to test their homes and apartments for the radioactive substance. According to officials, the potential for radon contamination is so widespread that the only people who do not need to test their homes are those living in apartments above the second floor.

Every detached house, row house or town house, mobile home with a permanent foundation, and apartment building should be tested, federal health officials declared. "There has been a very passive mood up to now," Margo Oge, the director of the EPA Radon Office, said. "Radon is something you cannot see or smell or feel, making it easy to ignore."

Radon contamination became a concern in 1984 when it was found in homes in eastern Pennsylvania, northern New Jersey, and New York, all of which are along a geological formation known as the Reading Prong. Staggering levels of contamination were found in these areas, with some measuring thousands of picocuries.

In 1987, the EPA surveyed homes in ten states and found health-threatening levels in one of every five homes. A year later, 7 more states were surveyed, and the EPA said more than 3 million homes in the 17 states are at life-threatening levels.

Federal guidelines say if the radon level hits four picocuries or above, the residence should be monitored. If that reading persists, then steps must be taken to get rid of the gas. When radon contamination reaches 20 picocuries, health officials recommend immediate action be taken to get rid of the gas. (See FIG. 1-2).

Radon is a problem in many countries. Europe recognized the deadliness of the invisible killer more than a decade ago. A 1979 study by the Swedish government determined that up to 40 percent of all cancer cases it examined were related to radon exposure!

SMOKERS VERSUS NONSMOKERS

Smoking is the number one cause of lung cancer, with radon ranking second. Studies indicate that smokers may be much more prone to getting lung cancer from high levels of radon gas than nonsmokers.

"While the relative risks for smokers and nonsmokers are comparable at lower exposures, the background lung cancer risk is much greater in smokers than nonsmokers," Mr. Jonathan M. Samet, M.D. said at an EPA symposium on radon. "For the reference category of no additional exposure, the lifetime risk of lung cancer for a male nonsmoker is 1.1 percent, whereas that for a male smoker is 12.3 percent. Radon's effects on lung cancer mortality are similar for females."

Every state has some high indoor radon levels, Richard Guimond of the EPA says. One of every three houses in the nation may be hazardous to the people living in them. One mother who had her house tested and discovered a high radon level told the news media recently, "My children were exposed to the equivalent of each smoking 23 packs of cigarettes a day for the first nine years of their lives."

The first major health hazard case to come to the attention of the nation was in 1984. A nuclear plant worker named Stanley Watras tripped radiation detectors at a plant near Reading, PA. The plant was not operating, and since Watras tripped the detectors

LUNG CANCER DEATHS ASSOCIATED WITH RADON

WL = 0.02
pCi/l = 4

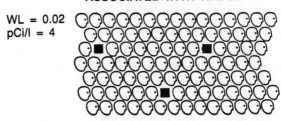

Between 1 and 5 out of 100

Fig. 1-2. The first three figures assume that these 100 people spent 75 percent of their time in the structure for 70 years, but they just as well could have lived in various dwellings with the same radon levels.

WL = 0.1
pCi/l = 20

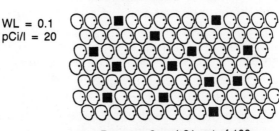

Between 6 and 21 out of 100

WL = 1.0
pCi/l = 200

Between 44 and 77 out of 100

If these same 100 individuals had lived only 10 years (instead of 70) in houses with radon levels of about 1.0 WL, the number of lung cancer deaths expected would be:

WL = 1.0
pCi/l = 200

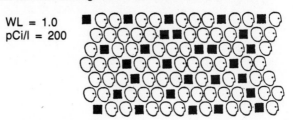

Between 14 and 42 out of 100

when entering, it was clear that he hadn't been exposed at work.

Plant officials took some measurements at the Watras home and found radiation levels that were 200,000 times above the level permissible for people living next to nuclear power plants. Watras and his family had been exposed to a radon equivalent of 455,000 chest X rays a year! Radon removal measures were taken, and the Watras house is now virtually radon free.

Most radon is breathed in and breathed out without change. The cancer risk comes from the successor atoms from radon decay, which are also radioactive. These are formed in the lungs or carried there on dust and smoke particles. Four picocuries represents as much risk as smoking eight cigarettes daily, the EPA estimates. Of every 1,000 people exposed to this level for a lifetime, 13 to 50 will get lung cancer that otherwise would not, the *Dallas Morning News* reported on August 5, 1987.

Naomi Harley, Professor of Environmental Medicine at New York University, says, "Nothing approaches the magnitude of this, not asbestos or formaldehyde. Radon is a major cancer causer and certainly leads the list of indoor pollutant problems in the nation." (See TABLE 1-1.)

Another way to put radon risks in perspective is to think about the risk associated with other activities. Table 1-1 gives an idea of how exposure to various radon levels over a lifetime compares to the risk of developing lung cancer from smoking and from chest X rays.

Table 1-1. Radon Risk Evaluation Chart

pCi/l	WL	Estimated number of lung cancer deaths due to radon exposure (out of 1000)	Comparable exposure levels	Comparable risk
200	1	440—770	1000 times average outdoor level	More than 60 times non-smoker risk
				4 pack-a-day smoker
100	0.5	270—630	100 times average indoor level	
				2000 chest x rays per year
40	0.2	120—380		
20	0.1	60—210	100 times average outdoor level	2 pack-a-day smoker
				1 pack-a-day smoker
10	0.05	30—120	10 times average indoor level	
				5 times non-smoker risk
4	0.02	13—50	10 times average outdoor level	200 chest x rays per year
2	0.01	7—30		
				Non-smoker risk of dying from lung cancer
1	0.005	3—13	Average indoor level	
0.2	0.001	1—3	Average outdoor level	20 chest x rays per year

2

Testing the Radon Level in Your House

In order to determine whether a house has elevated radon levels, measurements of radon or radon progeny in the house air must be made.

Measurement techniques are of two types: active and passive. Passive methods do not require pumps or specialized sampling equipment. Charcoal canisters and alpha-track detectors are convenient and relatively economical passive measurement devices. These simple-to-use methods also have the advantages of providing averaged, integrated measurements over a period of time, ranging from a few days for the charcoal to a few months for the alpha-track. Averaging over several months provides a much more meaningful measure of the concentration to which homeowners are being exposed. Time averaging is important because radon levels often vary significantly over the period of one day, as well as from season to season. (*Note:* For a Home Inspection Survey and an EPA Survey Form, see Appendix, "Compiling the Facts.")

ELECTRET-PASSIVE ENVIRONMENTAL RADON MONITOR

The EPA has recently issued a protocol for the use of a new passive radon measurement method. This new device, called an Electret-Passive Environmental Radon Monitor (E-PERM), is capable of making short- and long-term radon measurements. The device works on the same principle as the ionization chamber detector, which has been used as a radiation detector for many years. Although EPA's experience with this measurement device is limited, it does have attractive features. It is reported to provide good integrated measurements of radon with time exposures that can range from one day to one year. The results can be read in the field with a special surface potential voltmeter. It is said to be insensitive to relative humidity, which makes it a candidate for measuring on-site radon concentrations in soil gas.

Other measurement methods are also available. These active methods require an experienced sampling team with specialized equipment to visit the house. Although it is possible for one person to set up and operate this equipment, it is more typical for two or more people to be involved. Active methods include continuous monitoring, grab sampling, and use of a Radon Progeny Integrated Sampling Unit. Because of the need for special equipment and for a sampling team, these measurements are expensive and less commonly used for initial radon measurements in a house. They find greater application in premitigation diagnostic testing and in evaluation of the performance of installed radon reduction systems.

The EPA has issued protocols for making measurements in houses using alternative measurement methods, with the objective of determining occupant exposure. The EPA protocols recommend a two-step measurement strategy in which: 1. an initial screening measurement is made to provide a relatively quick and inexpensive indication of the potential radon/progeny levels in a house; and 2. additional follow-up measurements are recommended if the screening measurement is above about 4 pCi/l (about 0.02 working level). Persons making measurements are advised to apply the methods in a manner consistent with these protocols.

PASSIVE MEASURING

Two general types of passive measurement devices are currently in common use, with a third device gaining prominence:

1. The charcoal canister uses activated carbon in a small container to absorb radon.
2. The alpha-track detector consists of a container with a small piece of plastic sensitive to the alpha particles released by the radon and radon progeny. The user exposes the device in the house for a specified period and

then returns it to the laboratory for analysis. For both devices, the result is the radon gas concentration in pCi/l. Neither determines the concentration of radon progeny.

3. The E-PERM uses an electret to detect the ions generated by radon decay.

The Agency has also established a Radon Measurement Proficiency Program, enabling organizations that provide radon measurement services to voluntarily demonstrate their proficiency in making radon – radon progeny measurements. Lists of firms that have successfully demonstrated their proficiency under this program are published periodically.

SCREENING MEASUREMENTS

If no prior radon measurement has been made in the house, the initial measurement should be viewed as a screening measurement, and the exposure times for the devices should be:

‣ **Charcoal canister**—2 to 7 days, as specified by supplier
‣ **Alpha-track detector**—3 months to 1 year (or less, if specified by supplier)
‣ **E-PERM**—1 day to 1 year, depending on electret selected.

The objective of the screening measurement is to provide a quick and inexpensive indication of whether the house has the potential for causing high occupant exposures.

For the screening measurement, the device should be placed in the lowest livable space, such as the basement. Within that livable space, the device should be placed in the room expected to have the lowest ventilation rate. Livable space does not have to be finished or even be used as living space.

The devices should not be placed in sumps, or in small enclosed areas such as closets or cupboards. The objective is to measure the highest radon levels that might be expected anywhere in the livable part of the house. If low radon levels are found at the worst-case location, the house may be presumed to have low levels everywhere.

Screening measurements should be made under closed-house conditions: doors and windows should be closed except for normal entry and exit, along with minimum use of ventilation systems that mix indoor and outdoor air, such as attic and window fans. Closed-house conditions should also be maintained for 12 hours prior to beginning the screening measurement—if the measurement is shorter in duration than 72 hours. If possible, measurements should be made during cold weather, which usually corresponds to the highest radon levels. As above, the objective of maintaining these conditions is to obtain the highest expected radon measurement for the livable part of the house so that a low level measured under these conditions can be presumed to mean that the dwelling will likely remain at least as low under less challenging conditions.

FOLLOW-UP MEASUREMENTS

When selecting a measurement technique and a schedule for determining occupant exposure, be aware that radon levels in a given house can vary significantly over time. While the magnitude of this variation is house-dependent, it is not uncommon to see concentrations in a dwelling vary by a factor of two to three or more over a one-day period, even when the occupant has not done anything that might be expected to affect the levels. Seasonal variations can be even more significant. In some houses, the daily and seasonal variations will not be this great. If a meaningful measure of the occupants' exposure to radon is desired, it is best to obtain measurements over an extended period—three months to one year—and during different seasons. Since the highest levels in most climates are likely to occur during cold-weather periods, some measurements should be made during winter months.

If the screening result is greater than about 4 pCi/l, follow-up measurements should be made. If the screening measurement yields a result less than about 20 pCi/l, follow-up measurements should include:

- ▸ **Charcoal canister**—canister measurements made once every 3 months for 1 year, with each canister exposed for 2 to 7 days, as specified by supplier
- ▸ **Alpha-track detector**—alpha track device exposed for 12 months. This approach is preferred over the quarterly charcoal canister approach because the year-long alpha track measures for the entire year rather than just for 2- to 7-day periods, thus giving a more reliable measure of occupant exposure.
- ▸ **E-PERM**—exposed for 12 months

These measurements should be made in the actual living area on each floor of the house that is most frequently used as living space. Measurements should be made under normal living conditions, rather than the closed-house conditions recommended for screening. The year-long measurement period is suggested because the health risks at 20 pCi/l and less are felt to be sufficiently low that the homeowner can take time to make a good measurement of annual exposure before having to decide upon action to reduce the levels.

If the screening measurement yields a result greater than about 20 pCi/l, but not greater than about 200 pCi/l, follow-up measurements are again suggested for confirmation before taking remedial action. However, we recommend an expedited schedule for these measurements because of the higher risks associated with continued exposure to these higher levels. Follow-up measurements should be completed within several months after obtaining the screening result. Suggested follow-up measurements are:

▸ **Charcoal canister**—a one-time measurement on each floor having living space, under closed-house conditions (during the winter if possible), with exposure for 2 to 7 days.
▸ **Alpha-track detector**—a one-time measurement on each floor having living space, under closed-house conditions, with exposure for 3 months (or less, if specified by supplier).
▸ **E-PERM**—a one-time measurement on each floor having living space, under closed-house conditions, with exposure for 1 month.

If the screening measurement yields a result greater than about 200 pCi/l, the follow-up measurement should be expedited and conducted under closed-house conditions over a period of days or weeks; a 3-month alpha-track exposure might not be appropriate in this case. Short-term actions to reduce the radon levels should be considered as soon as possible. If this is not possible, it should be determined, in consultation with appropriate state or local health or radiation protection officials, whether temporary relocation is appropriate until the levels can be reduced.

In screening and follow-up measurements, the charcoal, alpha track, and E-PERM should be positioned within a room according to the following criteria:

▸ The device should be in a position where it will not be disturbed during the measurement period.

▸ It should not be placed in drafts caused by heating/air conditioning vents, or near windows, doors, or sources of excessive heat, stoves, fireplaces or strong sunlight.
▸ It should not be placed close to the outside walls of the house.
▸ It should be at least 8 inches (20 centimeters) below the ceiling and 20 inches (50 centimeters) above the floor, with the top face of charcoal canisters at least 4 inches (10 centimeters) away from other large objects that might impede air movement.

EPA ACTION LEVEL

The EPA has established an action level for indoor radon at 4 pCi/l as an annual average. This means that, while the radon concentration may fluctuate from day to day and season to season, its yearly average should not exceed 4 pCi/l. If the annual average does exceed 4 pCi/l, action should be taken to reduce the radon level. If such action is initiated, the objective should be to reduce the radon concentration to as low a level as is practical. The bulk of this document is intended to provide advice on reducing the indoor radon concentration.

This action level of 4 pCi/l does not imply that radon levels below 4 pCi/l are safe. Exposure to any measurable level of radon has an associated health risk. There are no absolutely safe levels of exposure. The individual must judge whether it is prudent to further reduce radon levels that are below 4 pCi/l. Note that with current technology it is not practical to reduce indoor radon levels below the local ambient values (typically 0.25 pCi/l).

At the opposite end of the spectrum, where radon concentrations are significantly higher than 4 pCi/l, urgency of the recommendation to reduce the radon concentration increases with the level of the radon. For high radon concentrations, temporary measures to reduce radon should be implemented.

In summary:

▸ For random concentrations greater than 200 pCi/l, action should be initiated within a few weeks.
▸ For radon concentrations in the range of 20 to 200 pCi/l, action should be initiated within several months.
▸ For radon concentrations in the range of 4 to 20 pCi/l, action should be initiated within a few years. The higher the radon the more urgent the need for action.

► For radon concentrations less than 4 pCi/l, no action is specifically recommended. However, many individuals may elect to further reduce radon concentrations in the range of 1 to 4 pCi/l.

CHARCOAL ANALYSIS

Evaluating the results is the essence of accuracy in any testing procedure. The laboratory that analyzes your charcoal test kits should have the finest equipment and trained technicians.

The Nucleus, Inc. located in Oak Ridge, TN—the capital city of radiation information/technology—is the number one supplier of counting systems for measuring radon concentrations in charcoal canisters. The highest standards of instrumentation are standard with the Nucleus equipment for radon testing professionals (FIG. 2-1).

The Nucleus system consists of the following components:

EN-26	2-inch Thick Lead Shield
3300	3-x-3-inch NaI (TI) Crystal Assembly
TS-2	Detector Base and Stand
5010	Amplifier/High Voltage Supply
PCA-1000	Pulse Height Analyzer Card for IBM PC
Ra-226	System Calibration Source
	Radon Application Software
	Electronic Scales

Radiation Accuracy

The NaI (TI) detector is housed inside a lead shield in order to prevent background radiation from the environment from entering the scintillation crystal. This greatly improves the sensitivity of the system and allows detector limits well below the current recommended level of 4 pCi/l of radon in indoor air. The pulse height analyzer is constructed on a card that fits inside the IBM PC or compatible. Software is supplied with the system to produce a live graphic display of the gamma ray spectrum in either color or monochrome on the computer monitor. This approach greatly simplifies the system operation. It is obviously apparent to the trained operator when the system is calibrated correctly, because a graphic display of the gamma radiation spectrum is produced for each charcoal canister as it is measured.

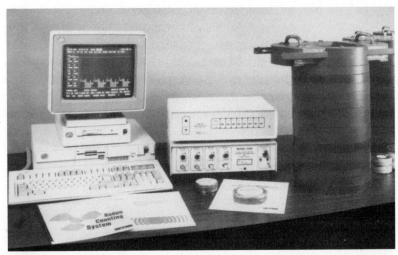

Fig. 2-1. The Nucleus Radon Counting System is said to be the most popular professional charcoal lab equipment.

Although the charcoal is quite efficient at absorbing radon gas, it also will absorb moisture from the air. This has the after-effect of reducing the amount of radon collected, and a correction must be applied during the analysis.

With the spectrometer built into the personal computer, an added advantage is realized: the corrections for exposure time, decay time, and moisture uptake may be applied automatically. The system from the Nucleus includes an extensive analysis software package. Correction tables for exposure, moisture, and decay have been programmed into the system, and the instrument will produce an analysis report showing the result in picocuries/liter with minimum operator intervention. Analysis of most common types of charcoal canister is possible without the necessity of running lengthy calibrations in known radon concentrations.

Radon Standards

With each system, the Nucleus also manufactures and supplies an NBS traceable radon standard. This calibration source is used to determine correct spectral alignment and absolute detector efficiency. It is also useful for conducting daily quality-control checks to ensure the system is operating correctly. Radon in the source container is generated continuously, and after one month, forms an equilibrium condition with its decay to produce a constant concentration. The calibration source will continue to generate radon gas for several thousand years and is very stable.

As the volume of business increases, the Nucleus system may readily be expanded to add more counting capacity without redundancy. Throughput is increased by adding more detector assemblies and electronics and multiplexing these into the one computer analyzer. Up to sixteen detectors may be interfaced to a single computer, but operator capability usually limits a single system to eight detectors (FIG. 2-2).

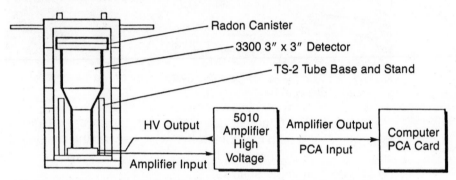

Fig. 2-2. The basic hookup diagram for a single detector in a Nucleus Radon Counting System charcoal test kit lab analysis operation.

The system offers additional flexibility to interface to database and software for custom reports and file handling. Operator training and assistance in proficiency testing are also provided as part of the extensive after sales customer support program. The basic system costs approximately $7,500. You can contact Nucleus, Inc. at P.O. Box 2561, Oak Ridge, TN 37831-2561.

RADON-ONE CHARCOAL MONITORS

Radon-One Lab Service has one of the finest charcoal test kits. It has passed Round Five of the U.S. Environmental Protection Agency's National Radon Measurement Proficiency Program, and its service is fast, concerned, and courteous (FIGS. 2-3 and 2-4).

Radon-One test kits are available for $15.95 from Radon-One Lab Service in Ohio. Send a check or money order made out to Radon-One to P.O. Box 205, Ridgecrest, NC 28770, and the test kit will be sent back to you quickly. All postage and laboratory fees are included in this price.

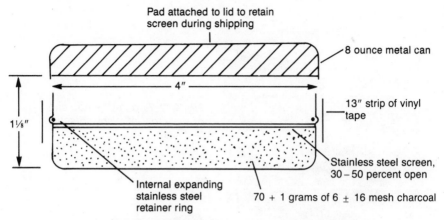

Fig. 2-3. *A diagram of a typical charcoal canister.*

TERRADEX ALPHA-TRACK DETECTOR KITS

Terradex Alpha-Track detectors are widely recognized for their quality. Their special price for inclusion in this book is just $19.95 for the indoor air detector. Checks or money orders should be made out to Terradex and sent to P.O. Box 206, Ridgecrest, NC 28770. This fee includes all postage and lab fees.

ELECTRETS

Although electrets may be an effective monitor type, the availability of E-PERMs is very limited. It is a new product, and it may be limited primarily to professional testing firms, and not readily available to homeowners. It seems the marketing of this monitor at this time is negligible. However, we will be happy to forward your inquiries to electret sources. Just send $2 for postage and handling to Memories, P.O. Box 155, Black Mountain, NC 28711.

CONTINUOUS RADON MONITORS

A continuous working radon monitor is *the* must-have appliance for people who want to know how much radon is in their homes at all times. Threshold Technical Products manufactures the Survivor 2 Home Radon Monitor. The high-tech device counts the radon level at all times (FIG. 2-5).

Pushing a button will reveal the average radon reading in the last 12 hours, and the full one-year operating mode will give you the average radon level in your home during the last 365 days. For those who want to know the radon concentration in their homes at

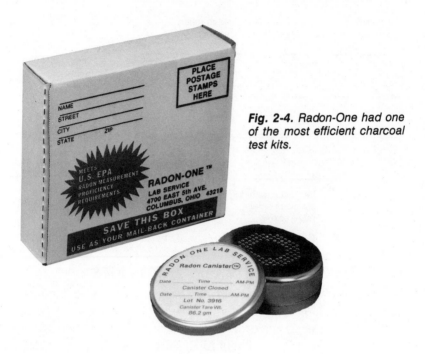

Fig. 2-4. Radon-One had one of the most efficient charcoal test kits.

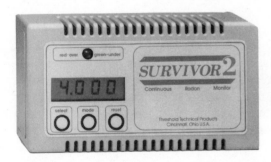

Fig. 2-5. Survivor 2 is the finest continuous radon monitor for homes and businesses.

any given time, this is the monitor to have. It notes both short-term and long-term levels, giving you a digital readout. Using an ion chamber detection method, the Survivor 2 has an accuracy of plus or minus 10 percent for an 84-hour period. It will measure radon from 1.2 pCi/l to 400 pCi/l. A green light notes when the level is below 4 pCi/l, and a red light burns and flashes when the radon concentration is more than that.

Survivor 2 Features

The Survivor 2 Home Radon Monitor weighs only 27 ounces, measures 6 inches long, less than 4 inches wide, and a little more than 2 inches thick. It plugs into any standard electrical outlet and is very simple to operate due to its microprocessor-controlled functions. The Threshold monitor may be moved around to various sections of the house and plugged into standard outlets. It is ideal to use after radon reduction methods have been taken, because a constant check may be kept on their effectiveness.

Because radon levels may change considerably within a few hours, a continuous monitor like the Survivor 2 keeps you informed of the fluctuations. The radiation instrument is a modern version of the meters and counters used by radiation specialists working in the first nuclear facilities. In an independent testing laboratory, the Survivor 2 was put into a radon chamber with an average reading of 11.23 pCi/l, and the monitor reading was 10.79, giving a plus/minus accuracy of 4 percent. This is extremely accurate for radon instrumentation.

The Radon Home Monitor sells for around $335, some $60 less than the Honeywell Home Monitor. You may order the Survivor 2 directly from Threshold Technical Products by sending a money order payable to that firm, to P.O. Box 155, Black Mountain, NC 28711. Visa or MasterCard credit cards are also acceptable. Send your credit card name, card number, and expiration date, along with your name, complete address, and daytime telephone number to the above address. Credit card orders may also be placed by calling toll-free, 1-800-458-4931, extension 7. In Ohio, call 513-530-5242, extension 7. The unit has a one-year warranty.

Long-term testing is recommended by the EPA, and a continuous radon monitor is the most inexpensive and convenient way to obtain the concentration levels. A very practical advantage of the Survivor 2-type monitor is that it can be moved around so easily. One monitor may be used to test many houses of friends and family over a period of time.

Tiny Geiger Counter

American Threshold Products also manufactures a miniature geiger counter that sounds an alarm if the background radiation exceeds a certain level. Think of this unit as you would a smoke detector. It merely detects a sudden radiation increase and alarms.

The mini-geiger counter was designed with those who live in radiation-prone areas of the nation, such as near Three Mile Island and other nuclear facilities, in mind. Some people are concerned about radioactive waste being transported by trucks on highways near their homes, and railroad train accidents that involve radioactive spills seem to be chronic.

To install this Survivor geiger counter-type unit, you merely plug it into any standard power outlet (FIG. 2-6). In case of a power failure, a battery takes over the operation. The radon alarm unit sells for $185, and there is a $25 bar-graph option. This is just a signal meter defining the direction from which any radiation is coming. You may order the unit in the same way and from the same address and toll-free number as the Survivor 2.

The geiger counter-type unit is also popular in Europe where the Chernobyl incident has been instilled in millions of minds. Swiss, German, and French models in 240 volts are popular in those countries.

The Survivor 2 is a superb continuous home and business radon monitor, while the geiger counter model is more of a radiation alarm.

Fig. 2-6. *A miniature geiger counter, this Survivor model is a radiation alarm only, ideal for those who live near nuclear facilities or fear radioactive truck and railroad accidents near their houses.*

TAMPER-RESISTANT MONITOR CAGES

Obviously, people do not want to buy houses with high radon levels, and real estate codes are already in effect in many areas. Making sure that radon test kits are not tampered with in any way, intentionally or accidentally, is something which must be considered.

There is a cage designed for real estate personnel, radon testing professionals, and home inspectors who want to ensure a radon test kit may be used with some security. The special box (FIG. 2-7) works with all charcoal, alpha-track, electret, and diffusion barrier type radon test devices. Insert the radon kit inside the cage, and place it on a level surface. Plug the power adaptor of the cage into an electrical outlet. Turn the switch and remove the key; the cage becomes activated, and an indicator light turns green. If the cage is tampered with in any way, the light turns red. Test kits should be used outside the cage also. Naturally, the results of the one in the cage should be the same as the one or more outside the

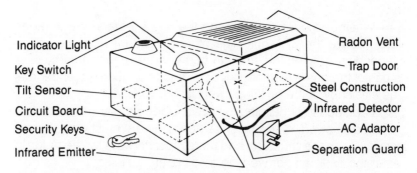

Fig. 2-7. The tamper-resistant cage for radon test kits is being widely used by real estate agencies to ensure radon readings have not been altered.

cage. Cages are also available which will accommodate continuous radon monitors, such as the Survivor 2.

You can purchase a tamper-resistant test kit cage for just $95 by sending a check or money order payable to Memories, P.O. Box 155, Black Mountain, NC 28711.

3

Radon Entry and Its Behavior

Almost every structure has some radon in it. The EPA estimates the average house in America has one picocurie of radon per liter of air.

Whether a building is among the 5 to 15 percent with radon levels above the EPA guideline of four picocuries per liter (pCi/l) depends mostly on the strength of the radon source nearby, how much of the gas is reaching indoors, and to a much lesser degree, the efficiency of the building's ventilation.

For a house or building to have a radon problem, three factors must be involved: radon must be nearby; there must be pathways for it to reach the structure; and there must be ways for the radon to enter the living and/or working areas of the building.

Typically most radon entering buildings comes from the soil or bedrock surrounding them or beneath them. Structures are really suction cups, pulling the gas through the soil into the interior. The difference between the air pressure outside and the air pressure inside a house sucks the gas inside. Houses have vents running from kitchens and baths, flues and vents running to exhaust heating systems and clothes dryers, and other passageways that serve to draw in the radon gas.

OTHER MEANS OF RADON ENTRY

Radon also gets inside buildings in three more ways. The gas diffuses into structures through holes and cracks in the foundation, and may even seep through the foundation material itself if it is porous.

Solid ground is not really so solid. Most soils measure 15 to 55 percent porous, which allows radon gas to move rather freely if there is a force to pull it along. "Solid as a rock" is not the most accurate cliché either. Weathered, fractured rock may come apart rather easily, because it may be webbed or splintered in its seeming solidarity. Shale and mineralized limestone formations, and even granite ground up by glaciers millions of years ago, also have sufficient pore space to allow gas to get through them.

Moist, humid ground slows down soil gas coming out of the dirt. This traps the radon gas, allowing it to build up concentrations.

Soil characteristics often differ considerably within a few feet of each other. It is certainly possible for two houses sitting beside each other—maybe 50 feet or less apart—to have totally different radon levels. One house may be on top of a rock formation with a very high radon concentration, but with a good layer of clay between the house and the radon. The clay is solidly packed and allows little or no radon to come through it.

The house next door is on top of the same rock formation loaded with radon, but it has very sandy soil between the foundation and the rock vein. The house sitting on the sandy soil base catches all the radon gas rapidly coming up through it. (See FIG. 3-1.)

AIR PRESSURE DIFFERENCES

As houses exhaust air, they pull in air to replace it. Air enters anywhere it can, through cracks and openings throughout the dwelling. This creates an inhaling negative pressure on the house, creating a suction on the ground on which it is built. Vents, chimneys, and appliances help draw in the outside air, but temperature differences between the outside air and the air inside the house also help.

Hot air rises because it is lighter than cold air. When it is colder outside than inside the home, the warmer indoor air goes toward the ceilings and floats on to the attic, the next floor, or outside, depending on cracks and openings. It is like a balloon losing pressure. The EPA calls this the "stack effect"; it is a vehicle on which radon gas can ride. Leaky ceilings, where the walls and ceilings may not be tightly sealed, crevices around chimneys, plumbing

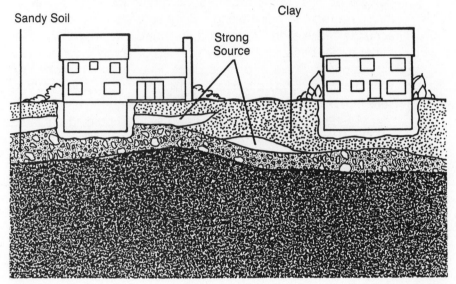

Fig. 3-1. Radon entry into individual homes depends on the impact of localized soil and geology.

chases (the vents and ducts, behind baths and showers), and ceiling fixtures allow air to pass. Ceiling fixtures might not be sealed tightly since building codes usually prohibit this. (See FIG. 3-2.)

The house starts to depressurize as the warmer air seeks to get out. If the temperature outside is 40 degrees colder than inside the house, the stack effect can make the air move at 200 cubic feet per minute (cfm) in an average-size house of standard construction. Depressurization may be decreased by sealing some of these openings, and sometimes this is done in conjunction with other radon mitigation methods.

Winter

The stack effect in winter is much greater, and this is why radon levels are much higher in homes in the winter than during the warm seasons. Houses are sealed as tightly as possible when it is cold outside. This traps the radon gas inside, and there are no open windows and doors to let the gas out. (See FIG. 3-3.)

NEGATIVE PRESSURE SOURCES
AND THERMAL BYPASSES

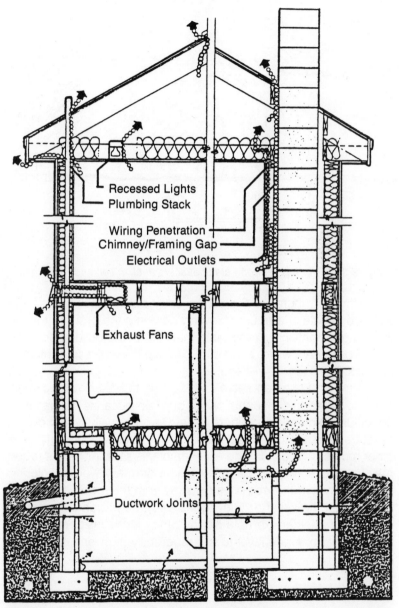

Recessed Lights
Plumbing Stack
Wiring Penetration
Chimney/Framing Gap
Electrical Outlets
Exhaust Fans
Ductwork Joints

Fig. 3-2. *Radon gas may even enter through electrical outlets and recessed lights.*

Fig. 3-3. *Winter has a strong stack effect to suck in soil and radon gas.*

Wind can increase the radon entry into houses also. Wind strikes one side of a house, blowing hard into the ground, forcing the soil-gas containing radon into that side of the building foundation or basement walls. Wind can increase the depressurization of a structure, as Fig. 3-4 illustrates.

Snow and rain can serve as a blanket to hold down soil gas. This allows radon concentrations to build up because the gas cannot diffuse into the atmosphere as fast as it normally would do. As moisture seals off the surface of the soil, this allows radon to seek other routes of diffusion. The essence of this situation is that radon may travel over longer distances than it normally would, entering houses maybe hundreds of yards away from the radon concentrations. When the water or snow goes away, the radon stops traveling. This is another example of why radon levels inside structures may vary from a safe level to a dangerously high one. (See FIG. 3-5.)

MAN-BUILT PATHWAYS

Inadvertently, man has built many roads to get radon inside his houses and buildings. In some cases, pipe lines and drain lines may be channels for radon entry. Many drainage systems

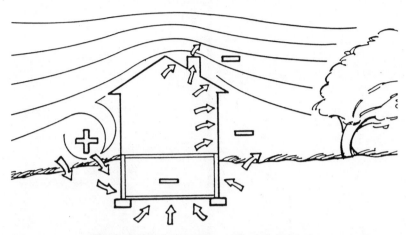

Fig. 3-4. Wind may drive radon into houses.

designed to prevent surface water problems may serve as passages for radon entry into buildings. Drainage pipes and perimeters around foundations sometimes are backfilled with sand or loose fill. These porous conditions entice radon entry.

RADON-CONTAMINATED MATERIALS

The third way radon gas gets into a house is when man has used some building materials loaded with radon during the construction process of the dwelling. This is rather rare, but it has happened hundreds, if not thousands, of times. In some Western states where uranium is mined, the tailings were once used in making some concrete products. These materials ended up in the walls and foundations of houses. As the radium decays, it releases radon gas into the house.

Fireplaces have been constructed out of radium-bearing stones, and the rocks used for thermal mass in some solar installations have been radon-laden. Even the old radium-painted clock dials, which glow in the dark, give off radon gas. Some people have rock collections in their homes or offices which produce the gas. Such things may not be primary sources of radon, but they may be contributing factors if the house already has a dangerous gas level.

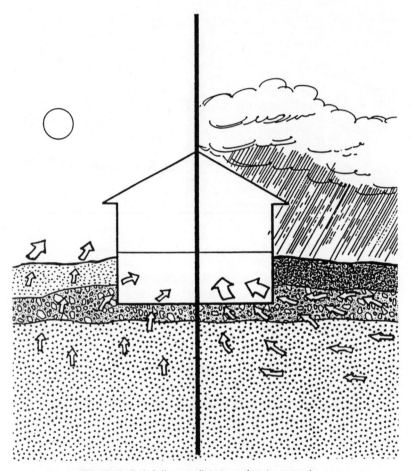

Fig. 3-5. Rainfall may dictate radon transport.

FOUNDATIONS

There are five different types of foundations popular in America. Most houses sit on piers, crawl spaces, slabs, full basements, or combinations of these. The section of the nation where a house is located usually dictates the type of foundation it has. Basements are much more common in the North and East while crawl spaces are much more frequent in the Southeast and Northwest. Slab-on-grade houses are by far the most popular in the West. Probably more than 95 percent of the houses in Florida are slab-on-grade.

Pier foundations, as illustrated in Fig. 3-6, are well ventilated and, by design, do not encourage radon entry. There is no pathway for pressure to build up and suck the soil gas into the building interior.

PIER FOUNDATION

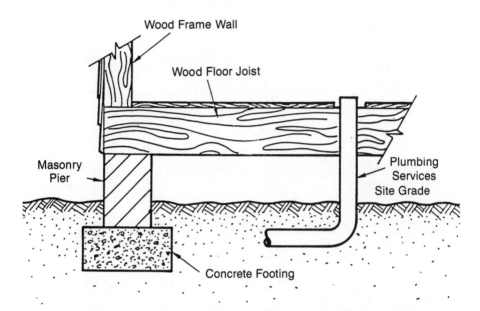

Fig. 3-6. *This is a typical pier foundation.*

Crawl spaces keep radon entry to a minimum potential, especially when there is a dirt floor and block walls. As the line drawing in Fig. 3-7 shows, crawl spaces may be ventilated or nonventilated. When crawl spaces are given no ventilation, radon levels may soar, since there is no outlet for the soil gas to escape. Floor or vapor barriers are recommended.

Basement foundations (FIG. 3-8) are much more prone to high radon levels, since such construction often creates a highly permeable zone surrounding a large below-grade section. A concrete floor, concrete block or poured walls, cracks between the walls and the floor, and plumbing penetrations are among the many things that may allow radon gas to enter if all are not totally secured to prevent radon entry. Trying to reduce the radon levels in homes with finished basements is much more difficult than those with other types of foundations.

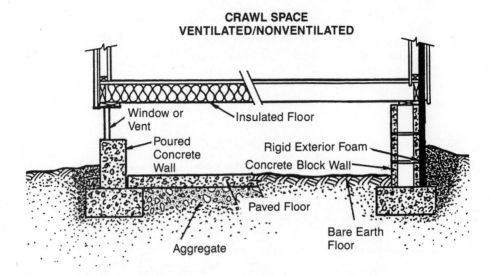

**CRAWL SPACE
VENTILATED/NONVENTILATED**

Window or Vent

Insulated Floor

Poured Concrete Wall

Rigid Exterior Foam

Concrete Block Wall

Paved Floor

Aggregate

Bare Earth Floor

Fig. 3-7. Crawl space characteristics are important considerations.

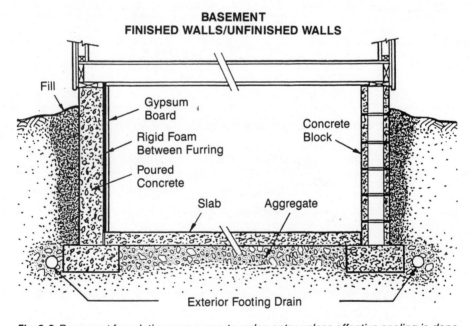

**BASEMENT
FINISHED WALLS/UNFINISHED WALLS**

Fill

Gypsum Board

Rigid Foam Between Furring

Poured Concrete

Slab

Aggregate

Concrete Block

Exterior Footing Drain

Fig. 3-8. Basement foundations are prone to radon entry unless effective sealing is done.

30

Slab-on-grade homes are also subject to dangerous radon levels. The concrete slab is usually perforated with plumbing pipes, which may allow radon gas to enter if they are not properly sealed and other measures are not taken (FIG. 3-9). Sealing openings between the foundation slab and the walls is very difficult in existing homes that may have dangerous radon concentrations.

SLAB-ON-GRADE WITH STEM WALL

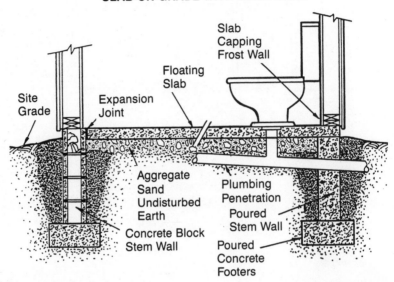

Fig. 3-9. *Basement construction is usually pierced in many places, may cause elevated radon levels.*

Combining foundation types may present multiple problems. When slab-on-grade sections are connected to basements and/or below-grade garages, elevated radon levels seem to be much more common. Figure 3-10 summarizes the foundation openings that may allow soil gas to bring radon inside.

The EPA list includes cracks in concrete floors and walls, sump holes that may be connected to earth, gravel drainage, drain pipes, waterlines, or other transport paths, and floor drains. Holes through slabs under tubs, showers, and toilets where space was left around traps let in radon. Holes in the slab around pipes, heating ducts run under slabs, and pores, cracks, gaps in mortar, and open tops in concrete block walls might be funnels for radon gas.

Slabs may be penetrated where wooden stair stringers or pieces of blocking have been inserted. Beware the cracks around support columns and the support walls for fireplaces. Floor-wall

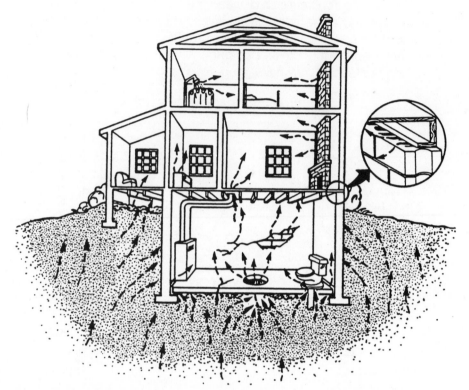

***Fig. 3-10.** Radon entry routes may be very numerous.*

joint cracks are common, as are those around water, gas, and utility line entry places.

INDOOR RADON CONCENTRATIONS

When radon gets inside a house, it can decay in the house air and continue decaying through the shorter-lived decay products, or it may be discharged outside with the air being ventilated before it has a chance to decay. Radon is often defined as having a half-life of 3.8 days. This means half the radon in the air has decayed during this period. Until the radon decay products (RDP) stick to something, they are referred to as the "unattached fraction."

These solid, electronically charged particles floating around in the air are the most dangerous. They get inhaled into the lungs. RDP's are equally hazardous to health when they get attached to dust, pollen, or tobacco smoke that is inhaled. When they stick to a wall, the floor, furniture, or other things, they present no danger. Of course, dusting and vacuuming may break them loose and put them back in the air where people may breathe them into the lungs.

Fans or blowers keep the air moving, including the radon decay products, and more of them end up being attached to the walls, floor, and furniture.

VENTILATION AND RADON

Radon readings indoors depend on the amount of radon coming into the building and how much dilution happens when the gas mixes with the house air and ventilated air. Ventilation efficiency is invariably higher during the summer. The EPA says there is some evidence to indicate air conditioning reduces soil gas entry in warmer climates. Radon readings during the summer and winter, according to a house study in Spokane, Washington, varied greatly.

How much air flows through a building is the *ventilation rate*, usually measured in cubic feet per minute (cfm), or air changes per hour (ACH). The number of times the air volume in a structure changes in an hour depends on the same factors that govern the airflow of soil gas in the ground. The mechanical equipment in the house, the difference between the temperature outside and inside, and the wind largely determine the ACH. Heating, ventilation, and air conditioning experts note the typical house in America built during the last 30 years has an effective leakage area (ELA) of between 1 and 2 square feet, resulting in an average infiltration rate of 0.4 to 0.1 ACH.

The more energy-efficient houses are much tighter and have 0.1 ACH and an ELA of 10 to 50 square inches. The greater the difference between the air pressures, the more increased the airflow through the house.

Although it has almost become a cliché that the tighter the house, the more concentration of radon, this is not really true. Several studies have been made concerning radon readings and air exchange rates in houses.

"None of the studies where ventilation rates were actually measured showed a correlation between tightness of house and elevated radon concentrations," the second edition of the EPA's *Reducing Radon in Structures* states. Houses with low ACH rates are *not* more prone to dangerous radon levels, studies document.

Deadly radon levels have been found in houses that were very drafty due to so many cracks and openings.

This does not say ventilation is not important! Effective ventilation is essential to the well-being of any house or building. A house or building must breathe. Cooking, washing, bathing, showering, and even overly misting the flowers are too much for a house that is not allowed to breathe. Even our own breathing adds almost two pounds of moisture to the air in the house daily. That is a mist you are exhaling right now.

Signs of Poor Ventilation

Buckling or curling shingles on a roof indicate trouble. Prolonged summer heat turns most attics into virtual ovens with temperatures of 140 degrees not uncommon. Hot spots form, and shingles start curling. In winter, water gets trapped under those same shingles. The water freezes, pushes the shingles up, and leaks ruin insulation.

Roof discoloration testifies to poor ventilation. Sick attics often force the tar out of shingles because of intense condensation and mildew.

Sweating walls, inside or outside, signify that more ventilation is needed. High temperatures, raining condensation, and trapped air push the paint off the eaves. Peeling paint elsewhere on the house denotes poor ventilation. Frosty, frozen, or foggy windows indicate an inadequate airflow. Icicle formations are caused by warm air coming through the insulation which melts the snow. As it runs off the roof, it refreezes. Notice that icicles tend to cluster in certain areas.

4

Removing Radon by Redirection

To get radon out of a structure, you suck it out or seal off its entry. Diverting soil gas from entering a building by depressurization is the most common mitigation means of blocking radon entry. Soil gas gets indoors primarily through the stack effect. The lower air pressure in the house allows the higher pressure in the soil to force the gas inside.

This suction on the soil may be reversed by using a fan to reduce the strength of the air pressure in the soil. Air is then forced out through any cracks or holes in the foundation, blowing the radon back down instead of sucking it into the house. Radon cannot get inside unless the soil gas carries it. Figure 4-1 illustrates a soil depressurization system that includes a pipe, fan, and an alarm system that goes off if the fan stops.

Reversing the stack effect with soil depressurization methods has been very effective in reducing high radon levels in a large proportion of homes and buildings.

DEPRESSURIZATION APPLICATIONS

The pipe and fan mitigation system has proven itself by depressurizing the soil and aggregates under concrete slab floors, concrete block basements and stem walls, cracks between floors and walls, under heavy vapor barriers, which cover bare earth floors in unfinished basements or crawl spaces, outside and inside footing drains, and sump holes.

GENERAL SOIL DEPRESSURIZATION

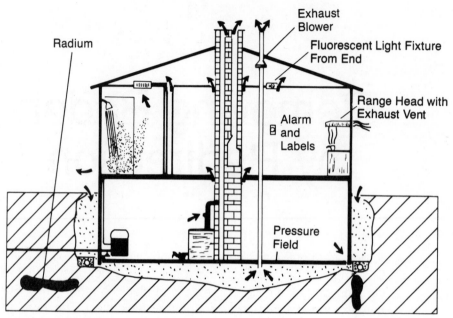

Fig. 4-1. *The soil depressurization system shown here features an alarm that will sound if the fan stops.*

"The key to successful soil depressurization is to extend the low pressure field around as much of the foundation as possible," the EPA says. "The field should extend under the entire floor and up the cores of the hollow block or up the outside of the foundation. This low pressure field needs to be stronger than the countervailing low interior pressures generated by the building in mid-winter."

It is easier to depressurize a house that is built over loose soil, has gravel under the concrete slab, and has a water drainage system, in which the concrete blocks have hollow cores and the soil has settled under the concrete slab, leaving pockets. These conditions make tapping into the radon much easier.

Generating a pressure field might be difficult where the soil is tightly packed, where slabs are setting on bedrock, and where the foundations have no drainage systems. Cored concrete blocks, stone walls, and footings under slabs are major barriers to the creation of a pressure field. Some soil permeability is so high—in an area around Spokane, Washington, for instance—it prohibits a pressure field.

SUB-SLAB SOIL VENTILATION

One or more pipes are inserted through the concrete slab into the ground, as Fig. 4-1 shows. The pipe may be inserted horizontally through a foundation wall at a level beneath the slab, as in Fig. 4-2.

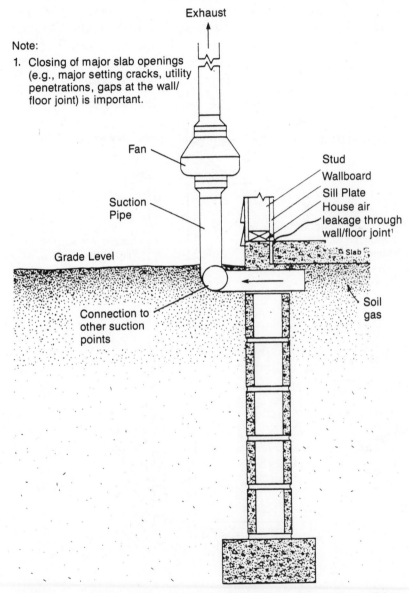

Note:

1. Closing of major slab openings (e.g., major setting cracks, utility penetrations, gaps at the wall/floor joint) is important.

Exhaust

Fan

Stud
Wallboard
Sill Plate
House air leakage through wall/floor joint[1]

Suction Pipe

Slab

Grade Level

Soil gas

Connection to other suction points

Fig. 4-2. Sub-slab suction, with pipes inserted through foundation wall from outside.

The mitigation fan really works well when the pipe is stuck into the gravel, which is often under the concrete in more modern houses.

Key factors concerning sub-slab ventilation include the number of ventilation points needed—usually one will suffice—where they must be placed, and the static pressure needed in the ventilation pipes to effectively keep radon from entering anywhere into the house. Figures 4-3 through 4-7 show five common sub-slab depressurizations. Radon researcher and authority Terry Brennan notes that system performance is often improved when a small hole—$1/2$ to 1 cubic foot—is dug under the suction point.

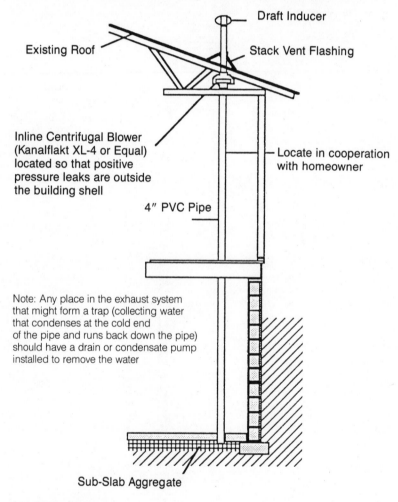

Draft Inducer

Existing Roof

Stack Vent Flashing

Inline Centrifugal Blower (Kanalflakt XL-4 or Equal) located so that positive pressure leaks are outside the building shell

Locate in cooperation with homeowner

4″ PVC Pipe

Note: Any place in the exhaust system that might form a trap (collecting water that condenses at the cold end of the pipe and runs back down the pipe) should have a drain or condensate pump installed to remove the water

Sub-Slab Aggregate

SOURCE: Brennan

Fig. 4-3. *Soil depressurization by suction on sub-slab.*

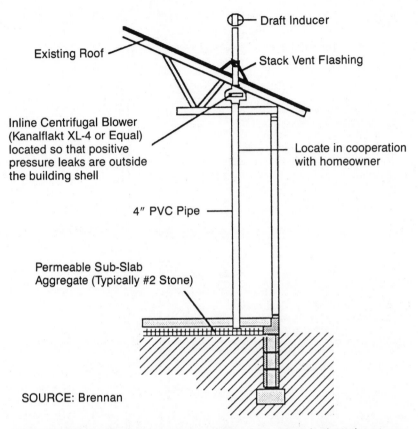

Draft Inducer

Existing Roof

Stack Vent Flashing

Inline Centrifugal Blower
(Kanalflakt XL-4 or Equal)
located so that positive
pressure leaks are outside
the building shell

Locate in cooperation
with homeowner

4″ PVC Pipe

Permeable Sub-Slab
Aggregate (Typically #2 Stone)

SOURCE: Brennan

Fig. 4-4. *Sub-slab suction on slab-on-grade. Any place in the exhaust system that might form a trap (collecting water that condenses at the cold end of the pipe and runs back down the pipe) should have a drain or condensate pump installed to remove the water.*

Some houses may not have any of these five construction details, but if they are built on permeable soil, sub-slab depressurization may still be employed by using more suction points. For very sandy or silt soil conditions, a pipe should be used for every 600 to 700 square feet of area along with a quality fan like a Kanalflakt.

"A 5-gallon-size pit should be routinely dug from under each suction point," the EPA says in *Reducing Radon in Structures*. "This usually increases the distribution of the pressure field and results in lower radon concentrations."

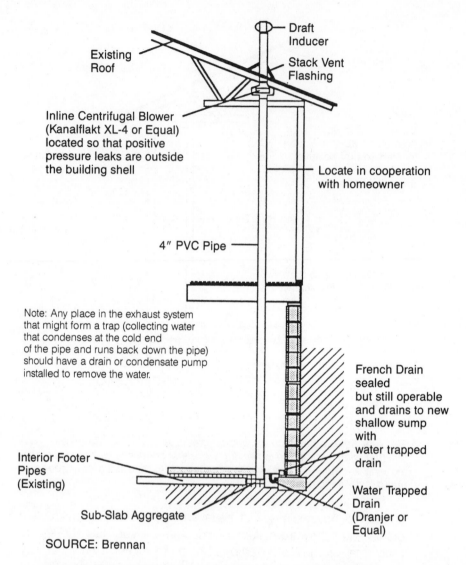

Draft Inducer

Existing Roof

Stack Vent Flashing

Inline Centrifugal Blower
(Kanalflakt XL-4 or Equal)
located so that positive
pressure leaks are outside
the building shell

Locate in cooperation
with homeowner

4″ PVC Pipe

Note: Any place in the exhaust system
that might form a trap (collecting water
that condenses at the cold end
of the pipe and runs back down the pipe)
should have a drain or condensate pump
installed to remove the water.

French Drain
sealed
but still operable
and drains to new
shallow sump
with
water trapped
drain

Interior Footer
Pipes
(Existing)

Water Trapped
Drain
(Dranjer or
Equal)

Sub-Slab Aggregate

SOURCE: Brennan

Fig. 4-5. *Soil depressurization by suction on sub-slab aggregate.*

Finding Suction Points

When trying to decide where to penetrate a concrete slab to locate
the pipes, you must be careful not to hit plumbing pipes, under-
ground electric lines, or radiant heat systems that may be buried in
the floor. Tapping on the floor may reveal hollow sounds where the
soil has settled, leaving holes.

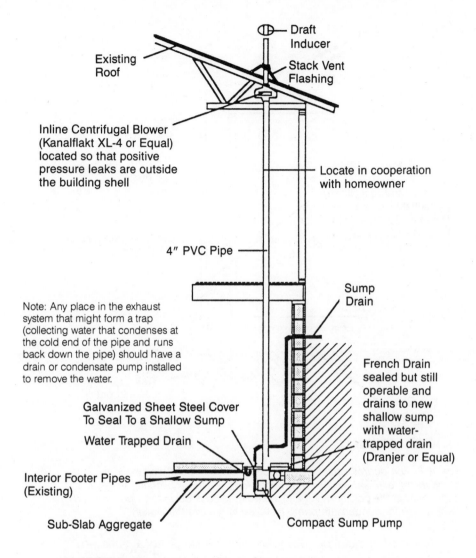

Draft Inducer

Existing Roof

Stack Vent Flashing

Inline Centrifugal Blower
(Kanalflakt XL-4 or Equal)
located so that positive
pressure leaks are outside
the building shell

Locate in cooperation
with homeowner

4″ PVC Pipe

Sump Drain

Note: Any place in the exhaust
system that might form a trap
(collecting water that condenses at
the cold end of the pipe and runs
back down the pipe) should have a
drain or condensate pump installed
to remove the water.

French Drain
sealed but still
operable and
drains to new
shallow sump
with water-
trapped drain
(Dranjer or Equal)

Galvanized Sheet Steel Cover
To Seal To a Shallow Sump

Water Trapped Drain

Interior Footer Pipes
(Existing)

Sub-Slab Aggregate

Compact Sump Pump

SOURCE: Brennan

Fig. 4-6. Soil depressurization by suction on sump hole.

Installing the Pipe

Almost all sub-slab suction systems use 4-inch PVC pipe running
from the 4-inch holes drilled through the concrete. A smaller size
may be substituted when the soil is very permeable. In some gravelly, sandy, or loose soils, 3-inch pipe is sufficient, while pipe only
1½ inches has been used in some sections of Florida where fine
sand is common.

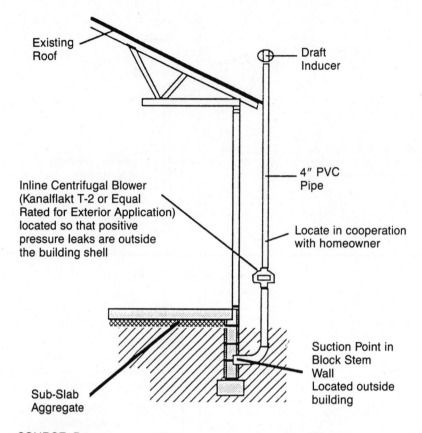

Existing Roof

Draft Inducer

4" PVC Pipe

Inline Centrifugal Blower
(Kanalflakt T-2 or Equal
Rated for Exterior Application)
located so that positive
pressure leaks are outside
the building shell

Locate in cooperation
with homeowner

Suction Point in
Block Stem
Wall
Located outside
building

Sub-Slab
Aggregate

SOURCE: Brennan

Fig. 4-7. *Soil depressurization by suction through stem wall exterior. (Any place in the exhaust system that might form a trap—collecting water that condenses at the çold end of the pipe and runs back down the pipe—should have a drain or condensate pump installed to remove the water.)*

The pipe should be wrapped with insulation to absorb any noise and to prevent water condensing on the outside of the pipe. When pipes have to be run through the house, they might pose unsightly problems. Decorative columns may be put around them, or they may be encased, in some instances, in attractive metal foil wrappings. Each instance dictates its own interior decorating challenges when the pipe has to run through formal rooms.

If a house must have more than one suction point, people often elect to run the pipes and holes against opposite foundation walls. If necessary, an exhaust manifold system may be designed where a lateral pipe connects the pipes. It is easier to run the pipes in houses with basements, but the upper floor, or floors, might pose more appearance problems.

Slab-on-grade houses cannot accommodate running a pipe network. People usually try to locate a suction point under a closet that hides the pipe. It runs on to the attic, where pipes can be connected to the fan assembly, and then run out the roof. If the pipe inside the house is going to cause difficulties, it might be possible to run the pipe on the outside of the house, as illustrated in Fig. 4-7.

SEALING

An array of sealants are on the market. Some quickly changed their packaging, claiming to be the best for radon reduction purposes, while new ones have appeared in an attempt to capture a big share of sealant sales.

Sikaflex Multi-Caulk made by the Sika Corporation is the unchallenged leader in quality. It is the only industrial-strength polyurethane multicaulk that has been selected for government buildings, resorts, and condos. Sika's sensational sealant has been around for a quarter of a century.

Since its invention in Switzerland in the 1950s, Sikaflex has sealed and caulked homes, skyscrapers, pleasure boats, pools, trucks, trains, trailers, buses, and ships of the U.S. Navy and Coast Guard, plus nuclear plants, industrial plants, and dams.

Sealing is essential to help ensure that sub-slab depressurization works well. Cracks where floors and walls meet must be sealed, especially French drains. Open sumps, daylight openings of footing drains, floor cracks, and all other openings through floors and walls must be sealed.

The Sika Corporation technical information sheet for the installation of Sikaflex Multi-Caulk for radon reduction sums up sealing:

Seal all openings, including pipes, drains, and electric conduits. Caulk completely around anything that protrudes through the walls or floors.

Perform perimeter caulking. Seal cracks between walls and floors, and all construction joints. The concrete slab should be a minimum of 30-days-old before caulking.

All cracks in walls and floors that are large enough to put a pencil point in must be caulked. Option A: Run a small bead of caulk over the crack. Then, with a putty knife, force caulking into the crack. Remove the excess caulk and recaulk with a ¼ inch of sealant on top of the crack, then tool. Option B: Using a hammer and chisel, vee the crack to approximately ¼ inch wide by ¼ inch deep, caulk, then tool.

Surface preparation is paramount. Proper cleaning of the cracks/joints is of utmost importance. With a wire brush, remove all loose paint, scale, rust, dust, and dirt on the surface to be caulked. Surfaces to be caulked must be dry.

Tooling is essential. Smoothing and finishing the sealant is absolutely necessary to ensure the sealant makes contact with the surfaces. Seal all cracks before tooling the Sikaflex Multi-Caulk. Tooling ensures excellent adhesion and the recommended shape factor. Dampening the tool—even if it is the back of a teaspoon—with warm, diluted soapy water helps to smooth the sealant.

The minimum thickness of the sealant in the cracks is 1/4 inch and should not exceed 1/2 inch. For cracks larger than 1/4 inch wide, you should consult Sika's Technical Service Department for detailed information at P.O. Box 297, Lyndhurst, NJ 07071. If immediate technical assistance is needed, call 201-933-8800. Sikaflex Multi-Caulk is available at most hardware stores and home centers throughout the nation.

Sikaflex polyurethane gives an excellent bond to concrete, masonry, steel, aluminum, copper, glass, ceramics, and wood. It performs well as an elastic adhesive between different materials, such as concrete to brick. It stays flexible and withstands natural expansion and contraction of materials without cracking. This makes it ideal for use around outside stairs, walks, basements, and foundation walls.

Multi-Caulk cures in approximately three hours and may be sanded, painted, or stained. Another asset of this polyurethane is that it does not sag down vertical surfaces, making it easy to apply to overhead areas. Radon mitigation professionals also like it because it guns easily. Low-odor Sikaflex not only seals, it conceals and comes in colors that complement homes.

Care should be taken to ensure the radon being sucked from the house to the outside does not have any way to get back into the house. If the fan and the exhaust system are properly placed, this will not happen.

DRAIN TILE SOIL VENTILATION

Perforated plastic or porous clay drain tiles surround part or all of some houses in the vicinity of the footings. These drain-pipe systems were installed to collect water and drain it away from the foundation. Drain tiles are generally located right beside or just above the perimeter footings, either on the side away from the house (often referred to as exterior drain tiles) or on the side under the house (interior drain tiles). When houses on slabs have drain tile under them, they are called "sub-slab" drains. (See FIG. 4-8.)

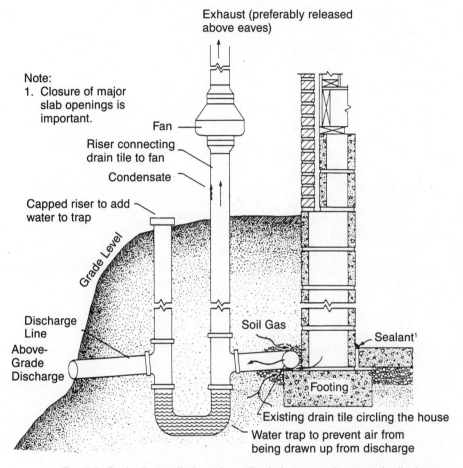

Fig. 4-8. *Drain tile ventilation where tile drains to an above-grade discharge.*

Water collected in the drain-tile system is usually routed to an above-grade discharge away from the house. This is easier when there is a slope at the site to carry the water to an area where it will be easily absorbed. Many houses have the water routed to a sump inside the basement or crawl space from which the water is pumped through a pipe to the outside, as in Fig. 4-9.

Drain tiles usually run right beside two of the major routes of soil gas entry. The first is the joint between the perimeter foundation

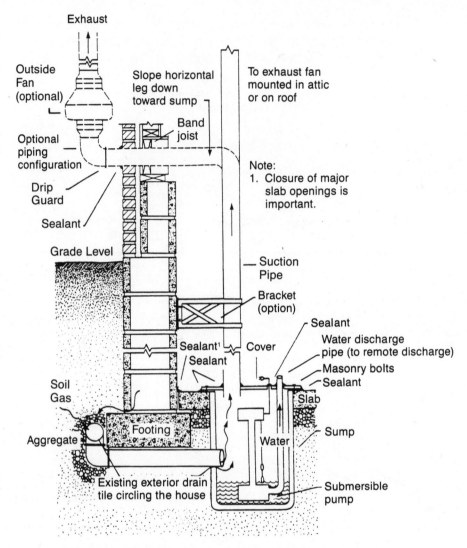

Fig. 4-9. *Drain tile ventilation where tile drains to sump.*

wall and the concrete slab inside the house. The other is the perimeter footing region where soil gas can enter the void network inside block foundation walls. Putting a suction on these drain tiles with a Kanalflakt fan can be effective in drawing soil gas away from these major routes.

The ideal conditions for drain tile depressurization systems include high permeability in the soil and crushed rock under the slab, and permeability in the soil under the footings and beside the foundation wall. When these conditions exist, there is high potential the drain tile suction can extract the radon from underneath the entire slab. Of course, the system works best when the house is completely encircled by drain tile. In some instances, drain tiles might be a convenient and in-place network for suction over a large area, but they must be in good condition, not broken and blocked with dirt. They should drain well without becoming flooded. Inspection of the ground around drain tiles might reveal wetness, indicating that they are not draining properly. Water staying in the drains reduces the fan's suction strength.

DIY Applications

Drain tile suction systems have reduced radon levels as much as 99 percent in some cases. The EPA says this can be used in houses requiring high degrees of reduction. If no drain tile network is in place, you can install one yourself, in some circumstances, for around $350. This might include the Kanalflakt fan, piping, and other materials. Since this system can be installed by the homeowner, it is especially practical where only limited degrees of reduction are needed.

DEPRESSURIZING BLOCK WALLS

Sub-slab suction is the preferred mitigation method of removing radon because it usually works well. If sub-slab suction cannot be used—or has not been very successful—depressurization of the block walls can be an effective alternative. Installing suction on walls from the outside is relatively simple, as illustrated in Figs. 4-10 through 4-12.

Inside basements, pipes may be run horizontally down the center of the building, so that suction is applied to the walls at each end. Lines may be run off to the sides to pick up the other two walls. Applying block wall suction from the outside keeps the system from interfering with inside space. The more simple the pipe installations, the more efficient suction may be.

Block wall suction effectiveness is totally dependent on sealing. Open block tops, plumbing and electrical penetrations, and all other openings must be sealed well. All walls might have to be covered with a sealant, due to the porosity of the concrete blocks and the number of small holes in mortar joints. This can be expensive, since quality sealants can cost more than 50 cents per square foot, and walls might range from 1,000 – 2,000 square feet.

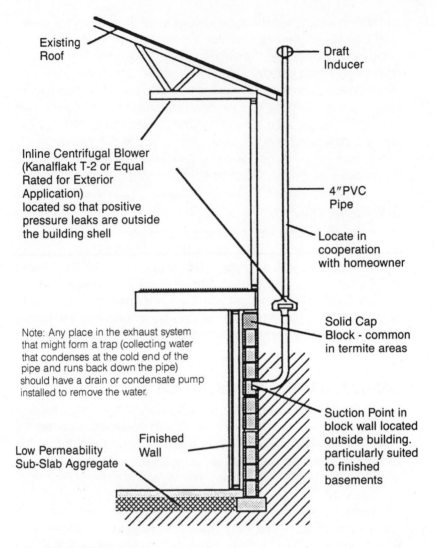

Existing Roof

Draft Inducer

Inline Centrifugal Blower
(Kanalflakt T-2 or Equal
Rated for Exterior
Application)
located so that positive
pressure leaks are outside
the building shell

4"PVC
Pipe

Locate in
cooperation
with homeowner

Note: Any place in the exhaust system
that might form a trap (collecting water
that condenses at the cold end of the
pipe and runs back down the pipe)
should have a drain or condensate pump
installed to remove the water.

Solid Cap
Block - common
in termite areas

Low Permeability
Sub-Slab Aggregate

Finished
Wall

Suction Point in
block wall located
outside building.
particularly suited
to finished
basements

SOURCE: Brennan

Fig. 4-10. Soil depressurization by suction of block wall exterior.

In some instances, masonry paints or stuccos may be used. The EPA recommends that you force some air into the basement when applying a sealant, since the air pressure might help push the sealant into the walls, leaving fewer pinholes. If the walls are often wet, they should not be coated until the water problem is solved.

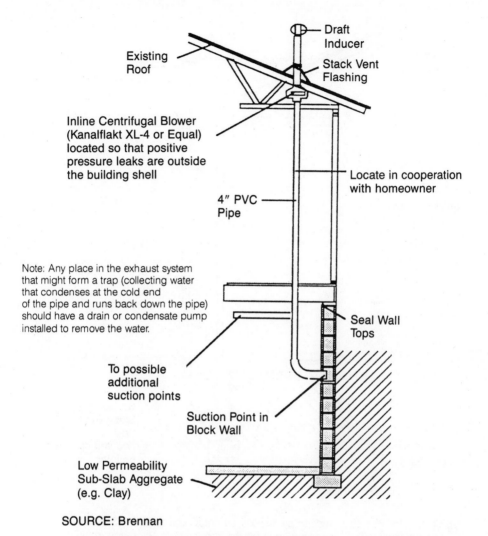

Draft Inducer

Existing Roof

Stack Vent Flashing

Inline Centrifugal Blower (Kanalflakt XL-4 or Equal) located so that positive pressure leaks are outside the building shell

Locate in cooperation with homeowner

4″ PVC Pipe

Note: Any place in the exhaust system that might form a trap (collecting water that condenses at the cold end of the pipe and runs back down the pipe) should have a drain or condensate pump installed to remove the water.

Seal Wall Tops

To possible additional suction points

Suction Point in Block Wall

Low Permeability Sub-Slab Aggregate (e.g. Clay)

SOURCE: Brennan

Fig. 4-11. *Soil depressurization by suction of block wall interior.*

BASEBOARD DEPRESSURIZATION

For homes without drain tile, a system is now available to help control water seepage and radon gas at the traditional point of entry: where the walls meet the floor. (See FIG. 4-13.)

A survey shows that more than half the basements in the nation have a water problem. Water seepage through walls and floors is unfortunately common, but the Beaver Basement Water Control System is an easily installed, economical solution. Beaver

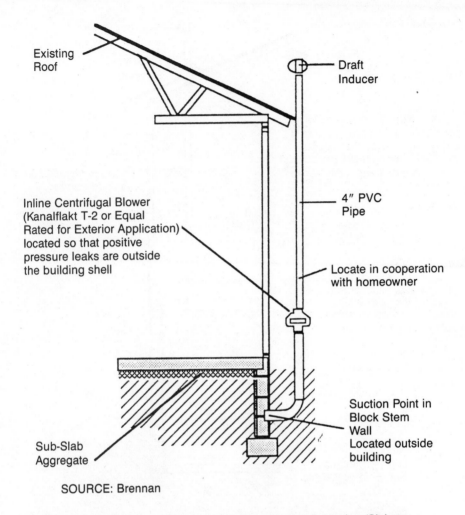

Existing Roof

Draft Inducer

4″ PVC Pipe

Inline Centrifugal Blower (Kanalflakt T-2 or Equal Rated for Exterior Application) located so that positive pressure leaks are outside the building shell

Locate in cooperation with homeowner

Suction Point in Block Stem Wall Located outside building

Sub-Slab Aggregate

SOURCE: Brennan

Fig. 4-12. *Soil depressurization by suction on stem wall exterior. (Slab-on-grade or bi-levels are good candidates for this technique.) Any place in the exhaust system that might form a trap—collecting water that condenses at the cold end of the pipe and runs back down the pipe—should have a drain or condensate pump installed to remove the water.*

Baseboard is a vinyl system installed around the perimeter of the basement.

Holes are drilled behind the baseboard into the hollow spaces within the blocks. The baseboard adheres to the floor and wall with a special two-component adhesive. It collects radon and gets rid of the water without having to undergo the expensive excavation to

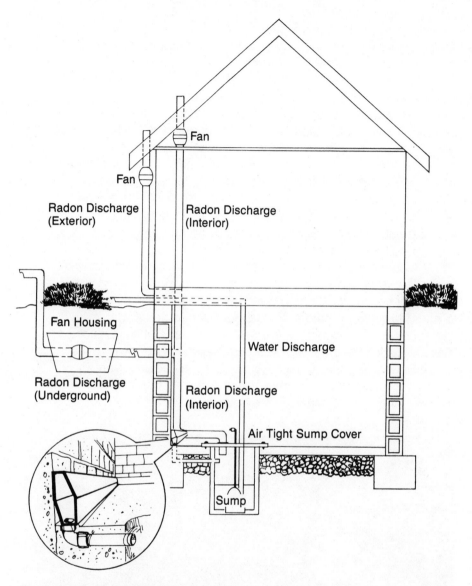

Fig. 4-13. *Beaver Baseboard is an effective and affordable radon- and water- control system alternative to traditional drain tile.*

build a drain-tile system. Proven to be effective in more than 30,000 homes, it is designed for do-it-yourself installation.

Because the system forms a water and airtight seal, the fan in the suction system can draw out the radon gas from the walls and under the slab, through the seam where the floors and walls meet.

For the Beaver radon-control baseboard to work, all holes must be sealed, especially the top of the blocks. Cracks and openings must be sealed, in the same way as the outside drain tile system. In some cases, radon levels have been lowered substantially with this type of system.

The Beaver baseboard utilizes the common crack or opening between the walls and floors, so be sure not to seal the crack prior to putting the baseboard in place. Most wet basement problems are caused by water collecting around the foundation, creating hydrostatic pressure. This pressure forces water through the joint where the floor and walls meet, and even through cracks and pores in the concrete block walls. The Beaver system quietly collects the water and drains it away to the sump pump. Figures 4-14 through 4-20 outline the installation steps.

Pressure is relieved by drilling holes in the concrete block cavities at floor level. It is easy to paint or panel a basement fitted with the Beaver system. It may be painted the day after installation. Prior to paneling, a vapor barrier should be fastened at the top of the wall and allowed to extend all the way to the floor. Cut it off at floor level and tuck it behind the baseboard. Nail furring strips over the plastic film barrier. A very neat appearance may be achieved with the Beaver baseboard. It is an excellent system to use in conjunction with other radon mitigation methods.

Fig. 4-14. *Step 1 for Beaver baseboard requires chiseling holes in the hollow spaces in the concrete blocks, but no holes are required in poured concrete walls.*

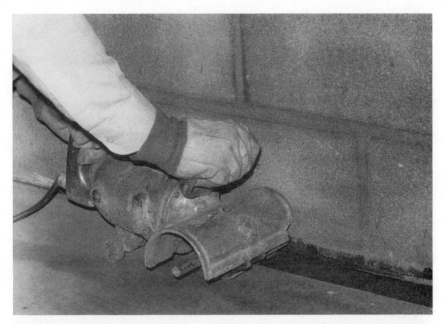

Fig. 4-15. *Step 2 for Beaver baseboard demands that the floor next to the wall be free of dirt, paint, or tile for at least 2 inches, essential to a watertight and airtight seal.*

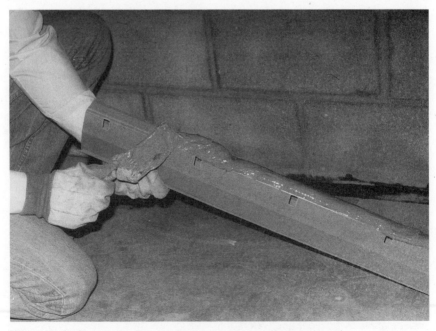

Fig. 4-16. *In Step 3, the Beaver Seal is applied to the baseboard.*

Fig. 4-17. In Step 4, the Beaver baseboard section is set in place.

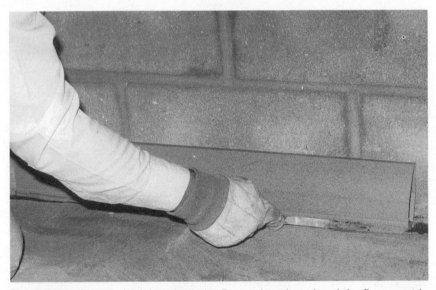

Fig. 4-18. In Step 5, the joint where the Beaver baseboard and the floor meet is sealed to make it airtight and watertight.

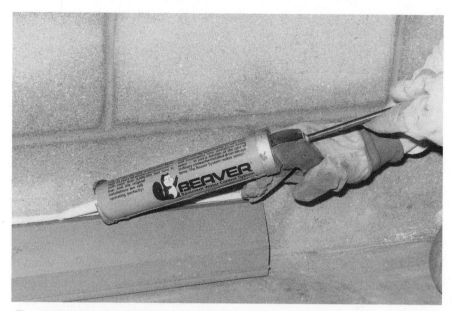

Fig. 4-19. *In Step 6, the top of the Beaver baseboard is sealed so that it is airtight.*

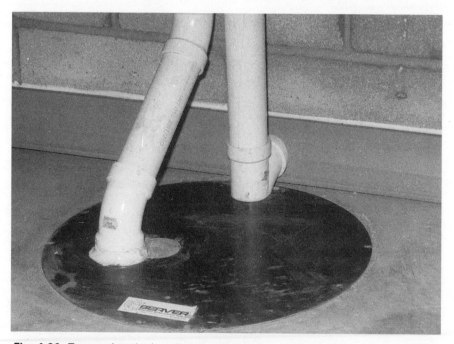

Fig. 4-20. *To complete the installation, a tight hole is cut in the Beaver baseboard, so that the PVC pipe may be inserted and then connected to the sump basin, which contains the sump pump.*

55

Beaver baseboard for water control and radon collection may be found in most hardwares and home centers. The Beaver Corporation is located at 890 Hersey Street, St. Paul, MN 55114. The phone number for technical advice is 612-644-9933.

Beaver baseboard may be used even if no water is present in basements. Instead of running the system to a sump pump, you can vent the system right through the roof, just like an interior pull on drain tile. Figure 4-21 is a line drawing from an EPA manual. It notes that the baseboard duct may be galvanized metal or PVC. Beaver baseboard is a strong vinyl that is simply a plastic cousin to PVC. Beaver baseboard should work equally well, if not better, than the other materials.

DEPRESSURIZATION UNDER PLASTIC FILMS

Many houses sit on crawl spaces that have earth floors, and some unfinished basements have dirt floors. A barrier is put down to cover the total bare-earth area. Ideally, it should be fastened well to the walls so that it will form as airtight a barrier as possible. Figure 4-22 shows a soil depressurization system under a heavy plastic film or vapor barrier.

BASEMENT PRESSURIZATION

Blowing air into the basement from the upper level is mentioned as a possible mitigation method in some EPA manuals. Air from upstairs is blown downstairs. This air pressure is supposed to prevent the soil gas containing radon to come up out of the ground. This approach requires sealing every possible crack and opening in the basement floors and walls. A tight basement is mandatory. If windows or doors are opened, the pressurization is lost.

This is more of a last resort if all other depressurization methods fail or are impossible to install. Pulling a lot of air from the living space and blowing it downstairs often causes ventilation problems. (See FIG. 4-23.) We do not recommend this method, but you can find information on it by contacting an EPA office.

As one EPA manual states, "Basement pressurization is not feasible in houses with woodstoves, fireplaces, or other vented combustion devices which could be backdrafted when air is drawn out of the upper floors."

SCHEMATIC OF SOIL DEPRESSURIZATION BY
SUCTION ON BASEBOARD DUCT

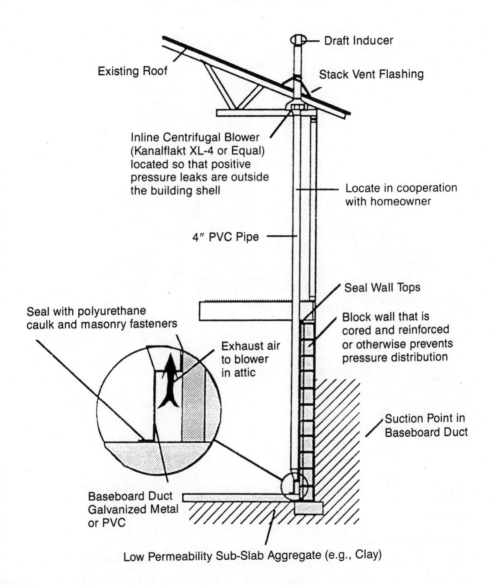

Draft Inducer

Existing Roof

Stack Vent Flashing

Inline Centrifugal Blower
(Kanalflakt XL-4 or Equal)
located so that positive
pressure leaks are outside
the building shell

Locate in cooperation
with homeowner

4″ PVC Pipe

Seal Wall Tops

Seal with polyurethane
caulk and masonry fasteners

Block wall that is
cored and reinforced
or otherwise prevents
pressure distribution

Exhaust air
to blower
in attic

Suction Point in
Baseboard Duct

Baseboard Duct
Galvanized Metal
or PVC

Low Permeability Sub-Slab Aggregate (e.g., Clay)

SOURCE: Brennan

Fig. 4-21. Suction may be applied to a baseboard duct to depressurize the soil.

SOIL DEPRESSURIZATION UNDER
HEAVY PLASTIC FILM - SCHEMATIC

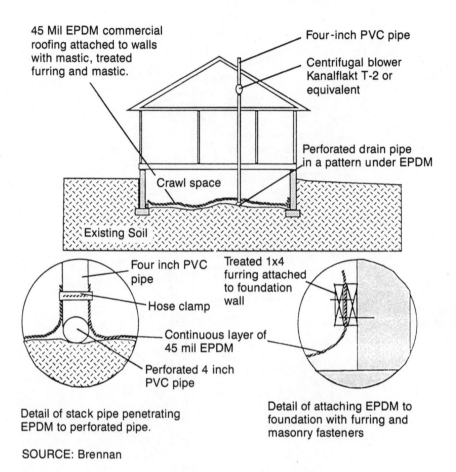

45 Mil EPDM commercial roofing attached to walls with mastic, treated furring and mastic.

Four-inch PVC pipe

Centrifugal blower Kanalflakt T-2 or equivalent

Perforated drain pipe in a pattern under EPDM

Crawl space

Existing Soil

Four inch PVC pipe

Hose clamp

Treated 1x4 furring attached to foundation wall

Continuous layer of 45 mil EPDM

Perforated 4 inch PVC pipe

Detail of stack pipe penetrating EPDM to perforated pipe.

Detail of attaching EPDM to foundation with furring and masonry fasteners

SOURCE: Brennan

Fig. 4-22. *PVC pipe may penetrate to the soil through a membrane to depressurize.*

Kanalflakt Fans

RB Kanalflakt, Inc. manufactures the finest high-performance, dependable, centrifugal in-line fans for radon mitigation methods. Virtually every EPA illustration notes the use of a Kanalflakt T1 Turbo 5 or T2 Turbo 6 fan—or its equivalent—in radon reduction methods. Some other fan brands use the same fan motor manufactured by Kanalflakt. Professional radon mitigation people use more Kanalflakt fans than all others combined.

BASEMENT PRESSURIZATION
USING UPSTAIRS AIR — SCHEMATIC

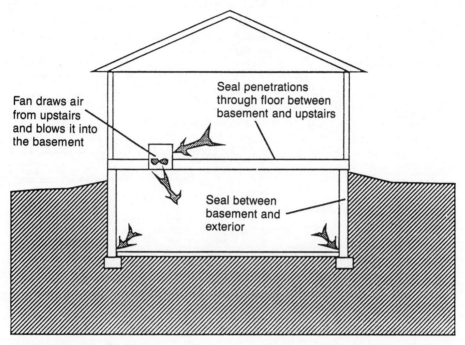

Fan draws air from upstairs and blows it into the basement

Seal penetrations through floor between basement and upstairs

Seal between basement and exterior

Fig. 4-23. Basement pressurization has had very limited usefulness because of a host of factors.

The two fan models are actually almost the same price, but they differ in applications. The T1 is smaller with a 1/40 HP motor while the T2 is larger with a 1/20 HP motor. The life expectancy of the fan running full time is more than 20 years. Powered by a revolutionary external rotor motor, the unit has a molded fiberglass case that is incredibly strong. It is impervious to most chemicals, weatherproof, corrosion-proof, and a new dimension in fan power and dependability. A T1 or T2 is excellent for sub-slab ventilation and for many other applications. Figure 4-24 shows a typical in-line installation, along with the measurements and performance and dimensional data on the Kanalflakt Turbo models.

Either the T1 or the T2 Kanalflakt models may be ordered through Memories (please make checks and money orders payable to *Memories*) at P.O. Box 206, Ridgecrest, NC 28770, for only $149 each. This is substantially less than the prices found at some radon product sources.

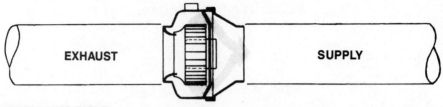

EXHAUST   SUPPLY

PERFORMANCE TABLE

MODEL	HP	FAN RPM	C.F.M. VS STATIC PRESSURE									DIA.	SONES
			0"	⅛"	¼"	⅜"	½"	¾"	1"	1½"	2"		
T1 Turbo 5	1/40	2800	158	143	125	114	90	45				5"	2.1
T2 Turbo 6	1/20	2150	270	255	235	200	180	140	110			6"	2.8
T3A Turbo 8	1/15	2150	410	375	340	285	225	180	135			8"	3.2
T3B Turbo 8	1/10	2300	520	500	470	445	415	310	230	200		8"	4.5
T4 Turbo 10	1/6	2400	700	670	640	612	582	470	410	250	115	10"	5.6
T5 Turbo 12*	1/8	1250	900	801	718	624	557	456	359	254		12"	4.5

* CONSULT FACTORY

DIMENSIONAL DATA

Model	Ø D	ø d	A	B	E	F
T1 Turbo 5	8	5	3.62	5.04	5.0	6.38
T2 Turbo 6	11½	6	3.54	5.51	6.97	6.38
T3A Turbo 8	13¼	8	3.58	5.79	7.95	6.38
T3B Turbo 8	13¼	8	3.58	5.79	7.95	6.38
T4 Turbo 10	13¼	10	3.46	6.18	7.95	6.38
T5 Turbo 12	16	12	4.33	8.27	9.25	6.38

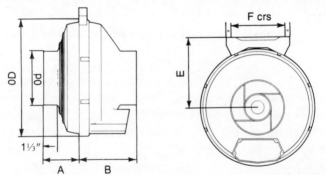

Fig. 4-24. Kanalflakt fans are considered one of the finest for radon reduction methods.

Kanalflakt centrifugal duct fans are ideal for many uses, as Fig. 4-25 notes.

SEALING STANDARD DRAINS

Basement floor drains and sumps are often holes to an unhealthy environment. Not only do these areas sometimes allow insects to enter your house, they tend to admit odors, promote mold spores, and may allow radon laden soil gas to come inside.

TYPICAL SYSTEMS

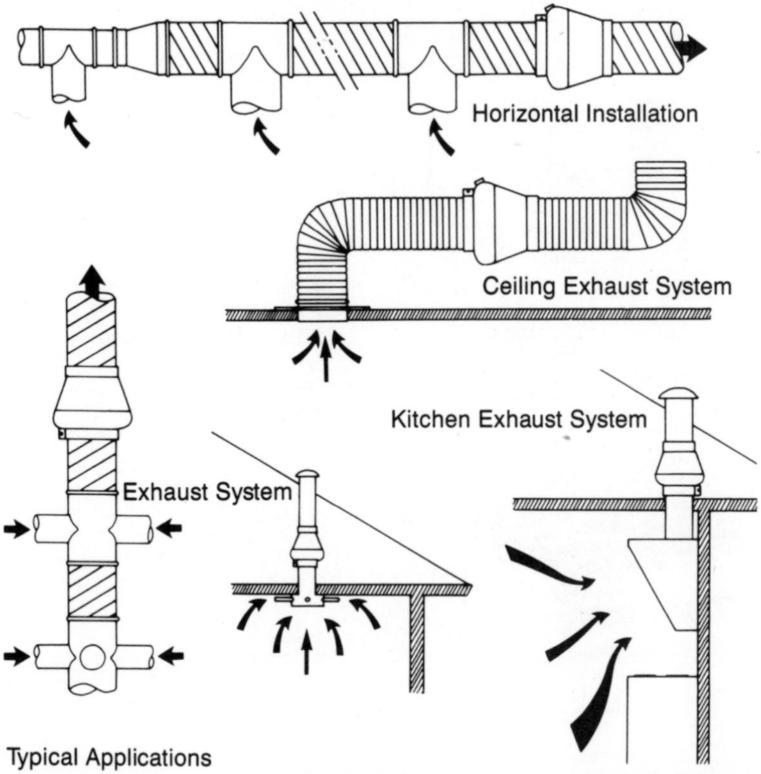

Horizontal Installation

Ceiling Exhaust System

Kitchen Exhaust System

Exhaust System

Typical Applications

Conference Rooms	Air Blast Heater	Soldering Booth Exhaust
Darkroom Exhaust	Plastic Extrusion Machinery	Welding Fume Exhaust
Toilet Exhaust	Projection Apparatus	Trailer Spray Painting
Locker Room Exhaust	Copying Machines	Manhole Ventilation
Kitchen Supply	Motor Cooling	Fume Control Systems
Duct Supply Booster	Pressure Booster	Oil Mist Exhaust
Air Curtain Supply	Computer Cooling	Portable Air Supply
Lab Hood Supply	Electrical Speed Control	Spot Ventilation
Trailer Refrigeration	Emergency Lab Exhaust	

Fig. 4-25. Kanalflakt fans may be used most effectively in many home and commercial situations.

EPA illustrations and literature often show Dranjer drain and sump seals. These seals allow water to continue to flow into drains or sumps, yet they help seal out radon gas. Inserting a Dranjer is very simple and takes only a few minutes. Quick and easy, it is one of the first steps in sealing off radon entry. Figure 4-26 outlines the insertion of the Dranjer that is most often used in homes to seal floor drains that range from 2 inches to 8 inches in diameter.

Allows water to enter your floor drain but seals out radon, insects, mold spores and obnoxious odors.

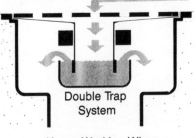

Double Trap System

Keeps Working When Trap Dries Out

Fig. 4-26. Dranjer model D-R2 fits most floor drains in basements, garages, and utility rooms that measure 2 to 8 inches in diameter.

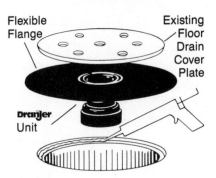

Flexible Flange

Existing Floor Drain Cover Plate

Dranjer Unit

Dranjers

Figure 4-27 has the schematics on Dranjer sump and floor seals. A Dranjer may also be attached to the underside of a sump basin, as in Fig. 4-28.

Dranjer model prices are $15.95 for the smaller-size D-R2 retrofit drain seal, $20.95 for the J-S5 sump model, $27.50 for the larger-size retrofit drain seal, and the J-N6 for a new installation floor drain in new construction is $29.95. All these models are available by sending a check made out to *Memories*, P.O. Box 206, Ridgecrest, NC 28770.

SUMP PUMPS

Sump pumps and sub-soil ventilation invariably work together, as most EPA literature and drawings indicate. Submersible sump pumps are ideal for finished basements where appearance is important, because they operate out of sight, below floor level.

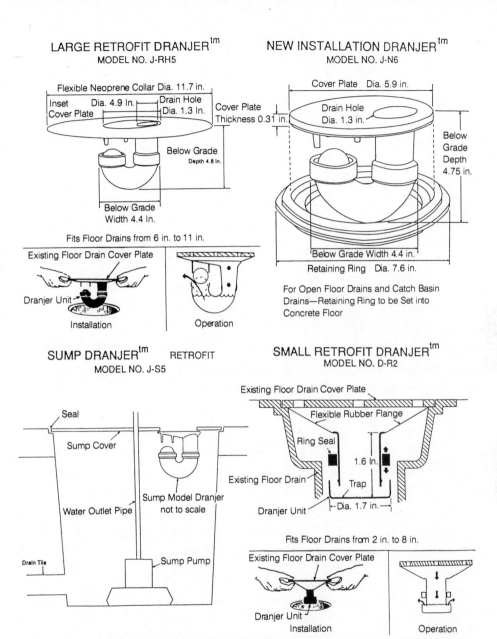

LARGE RETROFIT DRANJER™
MODEL NO. J-RH5

Flexible Neoprene Collar Dia. 11.7 in.

Inset Cover Plate
Dia. 4.9 In.
Drain Hole Dia. 1.3 In.

Cover Plate Thickness 0.31 in.

Below Grade Depth 4.8 In.

Below Grade Width 4.4 In.

Fits Floor Drains from 6 in. to 11 in.

Existing Floor Drain Cover Plate

Dranjer Unit

Installation

Operation

NEW INSTALLATION DRANJER™
MODEL NO. J-N6

Cover Plate Dia. 5.9 in.

Drain Hole Dia. 1.3 in.

Below Grade Depth 4.75 in.

Below Grade Width 4.4 in.

Retaining Ring Dia. 7.6 in.

For Open Floor Drains and Catch Basin Drains—Retaining Ring to be Set into Concrete Floor

SUMP DRANJER™   RETROFIT
MODEL NO. J-S5

Seal

Sump Cover

Sump Model Dranjer not to scale

Water Outlet Pipe

Drain Tile

Sump Pump

SMALL RETROFIT DRANJER™
MODEL NO. D-R2

Existing Floor Drain Cover Plate

Flexible Rubber Flange

Ring Seal

1.6 In.

Existing Floor Drain

Trap

Dranjer Unit

Dia. 1.7 in.

Fits Floor Drains from 2 in. to 8 in.

Existing Floor Drain Cover Plate

Dranjer Unit

Installation

Operation

Fig. 4-27. *Dranjer floor and sump seals accommodate most sizes to block radon entry.*

However, getting below floor level means getting closer to soil gas and radon.

Submersibles are also an advantage in households with small children because the pump and sump basin may be sealed

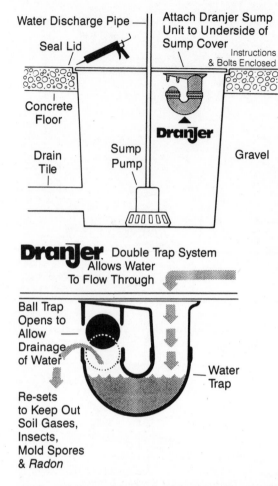

Water Discharge Pipe

Seal Lid

Attach Dranjer Sump Unit to Underside of Sump Cover

Instructions & Bolts Enclosed

Concrete Floor

Dranjer

Drain Tile

Sump Pump

Gravel

Fig. 4-28. *Dranjer model JS-5 seals sumps with a double-trap system that resets to keep out radon gas.*

Dranjer Double Trap System

Allows Water To Flow Through

Ball Trap Opens to Allow Drainage of Water

Water Trap

Re-sets to Keep Out Soil Gases, Insects, Mold Spores & *Radon*

against their entrance. In most applications, a submersible sump pump can replace a pedestal sump pump. Wayne sump pumps have long been known for their rugged construction, long life, and reliable performance. Plugging them into an outlet is just part of the easy installation. They are permanently lubricated and never need oiling. One Wayne model has a mercury switch design (FIG. 4-29), which provides maximum pump life. Another model has a reliable mechanical float switch design when compact size is a must.

Wayne sump pumps may be found in most major hardware and home center stores, including Ace Hardware and True Value Stores. The most popular models are the CDU-790, with a 1/3 HP made of cast iron, and the 1/2 HP CDU-800 model in cast iron. They are usually priced at $139.99 and $159.99, respectively.

Fig. 4-29. *The Wayne sump pump with a mechanical float switch is available in many hardware stores and home centers.*

Blue Angel Sump Pumps

Blue Angel sump pumps are a professional pump line. These high-quality pumps are very popular with radon mitigation professionals and are listed in many professional radon product catalogs. The 1/3 HP Blue Angel model BCS-33 turns on when the water level reaches 9 inches and turns off when the water gets down to 3 1/4 inches. It has a most dependable float switch, as well as automatic overload protection. It fits in sump basins 10 inches and wider.

The Blue Angel model BCS-50 is 1/2 HP with the same versatility, dependability, and economy. Both these Blue Angel professional sump pumps are available for $140 and $160, respectively, from Memories, P.O. Box 206, Ridgecrest, NC 28770. (Be sure to use a check or money order and include $5 for shipping east of the Mississippi or $10 for shipping west of the Mississippi.)

DIRECT VENT HEATING SYSTEMS

EPA manuals and literature concerning radon often refer to the problems of most central hot-air furnaces common to most homes. They tend to increase radon entry due to depressurization, because they suck air out of the area surrounding them for combustion and then create more stack effect when they send hot air and smoke up the chimney.

Basement air being drawn up through a heated chimney usually contributes much more to depressurization than combustion air consumption. Florida is one of the first states to require, in its guidelines for radon resistant construction, that all combustion heating systems be supplied with outside combustion air.

Apollo HydroHeat

Apollo HydroHeat is the most advanced, efficient, and economical of heating systems. It is a gas water heater that also heats the total house. For one to two hours daily, the typical residential water heater produces hot water used to wash clothes, dishes, and people. For the other 22 to 23 hours of the day, the hot water just sits there in the tank idle, doing nothing but maintaining its temperature, and if the tank is well insulated, that doesn't take any energy either.

Apollo's HydroHeat System takes hot water from a high-efficiency water heater and forces it through a specially designed water-to-air heat exchanger. The heated air moves gently into the home at a soothing comfortable temperature. The tank size assures there is always plenty of hot water for washing, bathing, and other household uses.

Apollo Systems have two-thirds fewer parts than heat pumps that are not nearly as efficient. Their simplicity of design and ease of installation make them superb to replace old, inefficient furnaces. HydroHeat units literally pay for themselves. It is much cheaper to heat water and homes with gas than it is electricity, as energy studies often report.

APOLLO COILS

The HydroHeat coil connects to a conventional or heating/cooling thermostat. When the thermostat calls for heat, the circulating pump on the coil begins circulating the warm water from the water heater through the finned tube heating coil. The blower has also been activated, and the air passing through the heating coil is heated and blown into the house through the duct system (FIG. 4-30). When the thermostat is satisfied, the pump and the blower both stop.

The air handler and the HydroHeat Gas Water Heater may be located as far away from each other as 70 feet! Depending on the house, condo, or apartment design, this can be a very important feature. And, the system does not need a chimney. It is vented

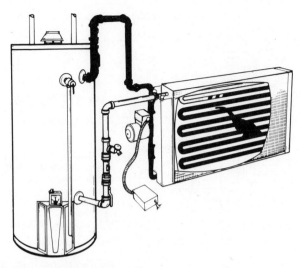

Fig. 4-30. *The Apollo gas water heater and coil are the primary components of the amazing heating system.*

through a wall directly to the outside. The unit is so compact and space saving, it may even be installed in a closet. The ease of installation is most appealing to homeowners and home builders.

MATCHED COMPONENTS

HydroHeat water heaters and air handlers must be matched according to the heating needs of the dwelling. The self-cleaning water heaters may be paired with air handlers of different capacities.

For the retro-fit market, the price range for a HydroHeat gas water heater and air handler runs from approximately $550 to $1200 plus installation costs. For a new house, the range is from $700 for the smallest set of matched components to $1400 for the largest system, plus installation costs. The Apollo Systems are the finest, and they are certainly the healthiest of heating and quality. Apollo is a division of State Industries, which manufactures the most elite water heater line for homes and commercial applications. Figure 4-31 shows a typical installation.

The Apollo HydroHeat & Cooling distributor in your area may be obtained by contacting Apollo at Cumberland Street in Ashland City, TN 37015. The phone number is 1-800-443-5793. Apollo also has cooling systems of many sizes for homes and businesses that are equally efficient and economical.

TYPICAL INSTALLATION

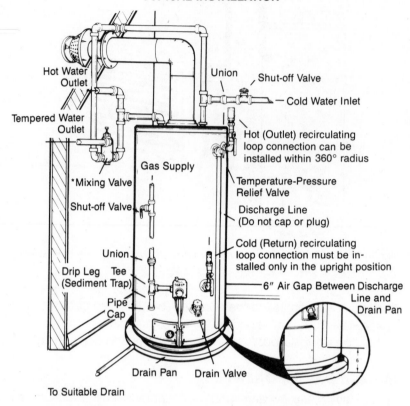

Fig. 4-31. *Apollo Heating Systems vent directly through a wall to the outside, requiring no chimney, and using outside air for combustion, reducing the stack effect for radon entry.*

The Apollo Direct Vent 5050 residential gas water heater is equipped with the exclusive, patented HydroSwirl dip tube, which creates a vortex of turbulence inside the tank, preventing any sediment accumulation. The unit cleans itself and stays energy efficient. No other gas water heater saves homeowners so much on operating costs. The EnergyGuide label found on all water heaters tells consumers how much it costs to heat water. And the HydroSwirl label tells a powerful story. The Direct Vent has the lowest operating costs of any gas water heater made!

DIRECT VENT FIREPLACE

Nine out of ten home builders and buyers say a fireplace is their favorite among the amenities. Superior Fireplace Company has a gas fireplace that does not need a chimney. It vents directly to the outside as the Apollo HydroHeat System does.

The attractive Superior Direct Vent Gas Fireplace may be installed in virtually any position along an outside wall, even under windows. With a thermal efficiency of 75 percent, it serves as a heater and fireplace in one. The unit may be installed in any room with an outside wall and on any floor, making it ideal for homes, condos, and high rise dwellings.

Since the direct vent fireplace needs no chimney and uses outside air for combustion, it does not add to any stack effect on the structure to suck in soil gas with radon in it (FIG. 4-32). The unit is excellent for room additions, home offices, and new construction. Its expansive view and flickering orange flames make it a focal point in any room.

The Superior model can be the only heat in the house in the event of a power failure! It has a self-generating millivolt control system with a piezo pilot ignition to ensure operation even if the electricity goes off. Natural gas and propane models are available.

Fig. 4-32. *The new Direct Vent Gas Model from Superior is the most healthy and attractive fireplace for installations where radon may be a concern.*

Superior is located at 4325 Artesia Avenue, Fullerton, CA 92633. Superior probably has the most extensive selection of fireplaces in the world, with designs appropriate for every decor and architectural demand.

5

Reducing Radon Levels by Dilution

Fresh air is a requirement for a healthy environment. Some level of air exchange is necessary to replenish oxygen and disperse contaminants produced in the home. Tightly sealed houses conserve energy and aid comfort, but they are often at cross purposes when it comes to a healthy indoor environment.

"Half of all the illness in the United States—including cancer and coronary, as well as respiratory diseases—are caused by the pollutants we breathe indoors," the U.S. Department of Health and Human Services said after a national health survey almost a decade ago!

The physiological basis for the impact of airborne chemicals on human well-being is the fact that humans are chemical beings. Motion, emotion, thought, speech—the human senses are all chemical processes. We can only get chemicals into the body through inhalation, ingestion, and absorption.

In their petition to the Consumer Product Safety Commission, the Consumer Federation of America, the largest consumer advocacy group, asked the CPSC to make Indoor Air Quality its number one priority. Indeed, the Indoor Radon Abatement Amendment was to the Toxic Substance Control Act!

NATURALLY POWERED VENTILATION

Opening windows and doors is the most basic type of ventilation. Letting outside air come inside and the indoor air go outside is the essence of fresh air, but temperature and weather conditions restrict this to almost nil. Even then, the EPA says windows should be opened in a particular pattern. In any event, naturally powered ventilation in modern homes is hardy effective.

PASSIVE VENTILATION

Crawl spaces in houses should follow Building Officials Code Agency (BOCA) guidelines, which advise 1 square foot of opening per 150 square feet of floor area without a ground cover. BOCA says a vent should also be placed within 6 feet of each corner. Vents should not be placed near plumbing pipes, which might freeze due to the air flow. If a crawl space has no vents, adding them generally helps. Passive venting in crawl spaces is best suited to houses with low-level radon readings (20 pCi/l) in mild winter climates. (See FIG. 5-1.)

Leslie-Locke makes an automatic foundation vent that opens at 70 degrees and closes at approximately 40 degrees. The foundation ventilator measures 16 inches by 8 inches, with a free area of 62 square inches. This static ventilator is used by many of the nation's quality builders, and it is often called for by architects designing homes and buildings.

The EPA does say passive ventilation of a crawl space can reduce radon concentrations. Although a ventilator is just a hole, it must perform three important functions: It must allow for the passage of air, it must hinder the entrance of the outside elements, and it must block insects from entering the home. Keeping these three duties in proportion is a challenge. If one function is changed, it affects one or both of the others. For example, if the air passage is enlarged too much, the weather protection decreases or the insect protection decreases. Covering the hole with grills or screens will restrict the size of the air passage.

ATTIC VENTILATION

The attic is the heart of a house, and nine out of ten older homes are gasping because they cannot get enough oxygen. Proper ventilation is essential to shelter, but it is one of the most misunderstood and misused concepts in modern construction. It is literally raining condensation in some attics right now.

When the world got out of bed this morning, millions of people took a shower while water was boiling in coffee pots. Some of the water from the shower turned to steam, just like the water on the

ISOLATING AND VENTILATING CRAWL SPACE—SCHEMATIC

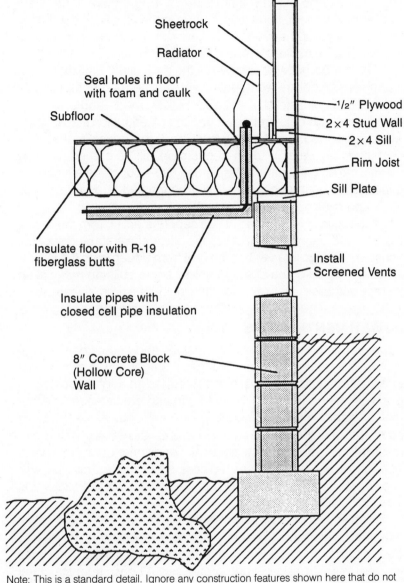

Note: This is a standard detail. Ignore any construction features shown here that do not appear in your house.

SOURCE: Brennan

Fig. 5-1. Passive crawl space venting is for low-level problems only in areas where winters are not severe.

stove. The water changed from a liquid to a gas. Hot air vapors rise, and they go right on through the ceiling if there is no vapor barrier to stop them. If the temperature in the attic is colder than in the living area of the house, the vapor converts back to a liquid when it hits the cooler air. And it "rains" in the attic.

The condensation compresses the thick, fluffy insulation, ruining millions of dollars worth each year, maybe even compacting it into a conductor instead of an insulator. The wood—and everything else—in the attic gets drenched, and the deterioration starts. Wood will rot if trapped in water long enough. If the house is effectively ventilated, the water in the gaseous state will be carried away by the moving air—the natural respiration rate—out of the attic before condensation can occur. Air should be inhaled through the soffit vents and exhaled through the roof ventilator.

If the attic is well sealed, it is a separate entity from the rest of the house. It cannot contribute to the stack effect of the rest of the house sucking soil gas with radon in it from underneath the house. Attics are not a source of radon unless some building material up there is outgassing radon, and that would be extremely rare.

POWER VENTILATORS

Power ventilators give an attic a breather, and Leslie-Locke manufactures the highest-quality line. It has the right power attic ventilator for every need, regardless of size or design. Eight models and two types range from 550 to 1,640 cfm and are available in either roof or gable mounted models. They help cut air-conditioning costs. Efficient power ventilation reduces attic heat, which can reach 140 degrees or more, so an air conditioner does not have to work as hard. The ventilator replaces super-heated attic air with cooler outside air. Leslie-Locke has the finest power ventilator line (see FIGS. 5-2 and 5-3).

High-efficiency, low-horsepower motors enable Leslie-Locke power attic ventilators to operate inexpensively. Most units use less electricity than two 150-watt light bulbs, and the ventilation insures your insulation investment and protects your house from structural damage due to water problems. Ridge vents are designed to provide positive air circulation for all ridge pitches. Leslie-Locke's Ridgerunner is the only complete unit for simple installation. This is considered the first major breakthrough in aluminum ridge ventilation in a generation.

Fig. 5-2. Leslie-Locke power ventilators include this model with a low profile and the highest efficiency.

Fig. 5-3. This cutaway of a Leslie-Locke power ventilator shows the fan design for maximum efficiency.

SOFFIT AND UNDEREAVE VENTS

In order to achieve maximum efficiency with power vents, ridge ventilators, and gable vents, it is necessary to provide ample intake ventilation. This can be easily accomplished by the installation of sufficient soffit under the eaves. To complete the Ridgerunner system, Leslie-Locke has a variety of quality under-eave louvers and vents. The giant ventilation corporation also has static roof louvers available in aluminum or ABS plastic.

Gable louvers are available from Leslie-Locke in adjustable- or fixed-pitch configurations. Midget louvers are recommended for moisture control in sidewall construction or other restricted areas. Leslie-Locke ventilation products are available in home centers throughout the nation. For the closest dealer to your area, contact Leslie-Locke at P.O. Box 723727, Atlanta, GA 30339.

Leslie-Locke Rotary Turbines

Turbine ventilators are the practical, proven, energy-free method to ventilate attics in homes and commercial buildings. Constructed of aluminum or galvanized steel, the ventilator is available in either externally or internally braced models, ranging in size from 6 inches up to 30 inches, as shown in Figs. 5-4 and 5-5.

Superior in air movement and with durable vane design, they offer a lower, less obtrusive roof profile. A new corrosion free, technologically advanced, Teflon-impregnated bearing system with stainless steel balls allows for trouble-free movement in the slightest breeze. The strong main shaft design adds strength in high winds. The 14-inch turbine/base combination has almost 40 percent more air movement than the standard 12-inch models. Both sizes have comprehensive warranties. They are guaranteed against defective materials and workmanship for the life of the house.

Turbine ventilators may also have a direct application to a passive radon reduction method. The Leslie-Locke turbine line is one of the most extensive in the world.

VENTILATION THROUGH A WARM-AIR FURNACE

It is possible to connect a 6-inch insulated pipe from the outside air to the return air duct on a central warm-air furnace system. When the distribution fan is not running, the air pressure differential between the outside and inside allows some air to enter. With the fan running, some 100 cfm of outside air is pulled inside. The

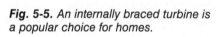

Fig. 5-4. *An externally braced Leslie-Locke Rotary Turbine, designed for high performance, durability, and value, is primarily used for commercial applications.*

Fig. 5-5. *An internally braced turbine is a popular choice for homes.*

makeup air from outside is heated only when the furnace is running. There is really no heat recovery. This is not a very effective system, most authorities say.

HEAT RECOVERY VENTILATORS

A heat recovery ventilator (HRV) is a package including fans, controls, and an air-to-air heat exchanger. Filtered fresh air is brought into the space from outdoors, and stale air is removed. In the process, heat—and sometimes moisture—is transferred from the warm to the colder airstream, saving heating costs in winter and cooling costs in summer. A quality HRV properly installed is considered one of the most effective means of radon reduction. It is also an excellent system for any dwelling since it keeps the indoor air fresh and removes a host of pollutants in addition to aiding radon reduction. Remember: no one particular method may be the total solution to reducing indoor radon concentrations, whatever the level; however, the HRV is a decided asset to any dwelling.

HRV Radon Reduction

HRV systems should be installed only by professional air-handling people. Although the systems are simple in design, they demand a lot of air handling details; the average DIY homeowner might not have sufficient knowledge or tools. There are three ventilation mechanisms that contribute to reduction of radon and the risk of lung cancer. Dilution is the most important mechanism. The radon level—or any other contaminant—is inversely proportional to the ventilation rate. This means that doubling the ventilation rate will halve the radon concentration. Four times the ventilation rate may reduce the radon level by 75 percent. These figures are based on perfect mixing, and the dilution or removal mechanism is enhanced when air is exhausted from areas of higher concentration.

The second mechanism, called "plate-out," occurs when radon progeny attach to particles that are attracted to surfaces in the home. Once attached to walls, furniture, and other things, they are no longer airborne and pose no threat to health. The circulation of air associated with ventilation enhances the plate-out process.

"Young air" is the third mechanism. This means that whenever ventilation is introduced, the individual atoms of radon spend, on the average, less time in the space. Because of the 3.8 days half-life of radon, any gas that is exhausted before it decays cannot give rise to alpha activity. The radon present yields less progeny and less of a working level because it is younger. Both the plate-out and young-air mechanisms of ventilation result in reductions in WL above and beyond the reductions measured in radon gas itself. So, it may be more meaningful to measure the impact of ventilation on the WL rather than, or in addition to, the radon level.

HRV Ventilation Efficiency

The layout of the ventilation system with respect to the space and the source will determine the ventilation efficiency. The entire area should be swept by the ventilated air. Introducing fresh air at one end of the basement and exhausting from the opposite end is one method. Efficiency may be improved further by exhausting close to a known area of higher radon concentration, such as a sump cover or French drain. Exhausting from the basement and supplying air to the first floor is sometimes useful because the basement has higher radon levels. In this upstairs/downstairs case, the entire basement is considered a source area, and the ventilation efficiency relative to the first floor is greatly enhanced.

Forced-Air System Interface

A standard dilution curve predicts the effect of increasing air-change rate on the radon concentration. The HRV is mounted to the basement ceiling. Stale air is taken from the basement or crawl space and ducted to the outside through a wall cap. A basement door undercut or grill must be provided to allow air supplied upstairs to return to the basement. Fresh air is brought from the outdoors through a basement-level wall cap. In the HRV, the fresh air is then conditioned by heat transfer—from the outgoing stale air—and ducted to the cold-air return of the existing forced-air system, and is then distributed throughout the house by the existing duct work (see FIG. 5-6).

When connected to a forced-air heating or cooling system, the air-to-air heat exchanger must operate continuously, or provisions must be made that will allow the heat exchanger to automatically operate when the heating or cooling circulation fan is operating. Continuous operation is recommended. The stale air exhaust and fresh-air intake caps must be a minimum of 6 feet apart. The air flow in the intake and exhaust air streams *must* be equal. This is accomplished by measuring air flows and placing a balancing damper in one or both airstreams. For radon applications, due to the inherent error in air flow measurement, systems should be balanced slightly positive, with 5 to 10 percent greater air flow in the supply stream.

Care must be taken not to introduce a negative pressure in the home, which could induce a higher rate-flow of radon-bearing soil gas·into the living space.

HRV in Basement and Living Area

Figure 5-7 shows an HRV installation in a basement. Figure 5-8 depicts an HRV installation in a ranch-style home on a slab. The HRV is mounted to the ceiling of a utility room or closet. Ducting must be insulated complete with a vapor barrier.

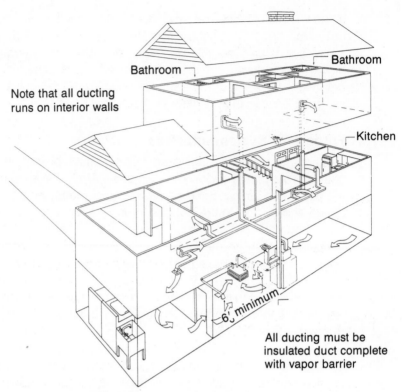

Bathroom

Bathroom

Note that all ducting
runs on interior walls

Kitchen

6' minimum

All ducting must be
insulated duct complete
with vapor barrier

Fig. 5-6. *AirXchange can use existing ducting in many houses effectively with one of its HRV models.*

A Quality HRV Line

AirXchange, Inc., 401 V.F.W. Drive, Rockland, MA 02370, has one of the most efficient and economical HRV lines. Its 502 Series is the health, comfort, and energy answer to removing indoor air pollutants, including radon, formaldehyde, excess humidity, odors, virus, bacteria, cigarette smoke, and other contaminants, replacing them with fresh, filtered outdoor air.

AirXchange's 502CA is a ducted, 70 to 200 cfm with a speed control, hanging straps, starting collars, and desiccant wheel. The suggested retail is $619, and it should be equipped with a frost control system which runs an additional $298 to $318. The 502 models economically provide more than one-half air change per hour in 1,000- to 3,000-square-foot houses, the minimum rate recommended by code authorities to ensure acceptable indoor air quality. (See FIG. 5-9.)

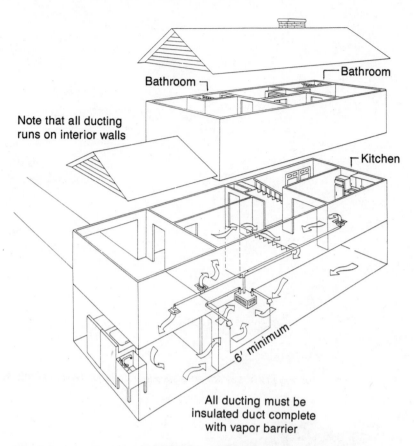

Bathroom

Bathroom

Note that all ducting
runs on interior walls

Kitchen

6' minimum

All ducting must be
insulated duct complete
with vapor barrier

Fig. 5-7. *An AirXchange basement installation is usually installed in a short time.*

The 570 HRV series includes a wall unit that is designed for 1,000-square-foot homes. The preferred location for it is an outside wall in a living or family room, centrally located for maximum ventilation efficiency. For fresh air while sleeping, the unit may be installed in a bedroom. The 570 sells for $498. The 570D is a ceiling unit designed for the same size house or area. It sells for $438 to $518, depending on options desired.

For larger homes and smaller business buildings, the Energy Recovery Module (ERM) sells for $1,995. It recovers up to 85 percent of energy required to heat and cool outdoor ventilation requirements. It reduces the cooling load up to four tons per 1,000 cfm.

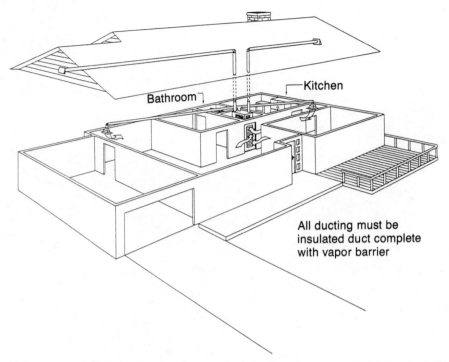

Fig. 5-8. An AirXchange HRV system on a slab foundation is typical of this schematic.

Bathroom⌐

⌐Kitchen

All ducting must be
insulated duct complete
with vapor barrier

Fig. 5-9. AirXchanges HRV's have one of the highest efficiency ratings and dependability records.

Rooftop HRV

The Model RTU 1000 Rooftop Energy Recovery Ventilator is a new
concept in rotary air-to-air heat exchangers. Designed as a pack-
aged unit for ease of installation and maintenance, it need only be
connected to ductwork and electrical power to make the system

operational. A unique rotary Energy Recovery Cassette slides in and out of the heavy duty blower cabinet. The self-contained RTU costs $2,863, plus the cost of a frost control, if the climate demands it.

HRV's literally pay for themselves, IAQ authorities declare. The ERM will usually pay for itself within two years, according to Bede Wellford or AirXchange. Contact AirXchange for information concerning the HRV most appropriate for your area and the dealer closest to you. Write to P.O. Box 155, Black Mountain, NC 28711.

6

Removing Radon by Filtration

Extensive research has been done seeking an air-cleaning device that will remove radon. As you are aware, scores of so-called different types of air cleaners are in a variety of stores. Almost every heating and cooling manufacturer seems to have an air cleaner.

The most effective filtration device is a portable fan-ion generator. Designed and developed by Ion Systems, Inc. in collaboration with a team of researchers at the Harvard School of Public Health, the patented system incorporates ionization, filtration, and air circulation. The small hassock-styled unit (FIG. 6-1) is highly effective in removing airborne radon decay products and in reducing the associated radiation dose to the lungs. Lab studies at Harvard showed this system of enhanced convection reduced the total potential alpha energy concentrations by as much as 95 percent, and reduced the mean dose to the bronchial tissues as much as 87 percent.

The portable fan-ion generator, named the No-Rad, yielded 75 to 90 percent effectiveness. More than a radon remover, it's also a highly effective ionizer, air cleaner, and air circulator. It helps cleanse the air of other pollutants, including dust, smoke, pollen, and bacteria, filtering out particles as small as $1/1000$ of a micron. When used as a fan, it helps cool a room in summer and helps circulate hot air during the winter.

No-Rad is much more effective than electrostatic cleaners, even when they have high-quality filters. Actually, electrostatic air

Fig. 6-1. *Sleek hassock styling makes the No-Rad unit attractive, and its light weight makes it easy to move to rooms where it is needed.*

cleaners can be dangerous when radon is present. True, the air cleaner will take dust particles from the air; as a result, the free floating radon atoms will have fewer dust particles on which to attach. The danger is, as studies document, that when free floating radon decay atoms are inhaled, the radiation to the lungs may be 40 times greater than that caused by radon decay atoms attached to dust particles.

IONIZATION AND RADON

The No-Rad's ionizer discharges both negative and positive ions into the air where they are widely circulated around the room by the fan. Highly charged ions act like magnets and attach themselves to the dust and radon decay products in the air. As the fan circulates the particles, they are either removed by the filters in the unit, or they come into contact with walls, ceilings, and room furnishings and adhere to them. Once collected on the filters or deposited on the surfaces—some of them can be knocked loose— they are no longer available for inhalation. The small amount of radiation that continues to be emitted from the collected or deposited particles is very limited, posing a low potential health risk.

Ions

Ions are electrically charged molecules that are constantly being created through the dynamic interaction of natural forces—rain, ocean surf, waterfalls, wind, and cosmic rays. Outside, ions bond together with air pollutants and settle out of the air to the ground.

Indoors, there is little opportunity to maintain or replenish the number of ions lost from the air. The No-Rad pulls pollutants from the air and charges the air, making it more like the fresh air outside.

Ion Systems, Inc. holds the worldwide exclusive license to design and manufacture a radon removal system based on proprietary technology developed, tested, and patented by Drs. Moeller, Maher, and Rudnick at the School of Public Health at Harvard University. Dr. Moeller is a member of the National Council of Radiation Protection and Measurement and chairs the NCRP Scientific Committee on Methods for the Control of Radon Inside Buildings. He has spent more than 20 years researching radon and its by-products.

No-Rad is effective up to 300 square feet, say a 15- by 20-foot room. However, the small unit may be easily moved to different rooms as needed. The unit is modestly priced at $299, and checks or money orders should be made out to Ion Systems, Inc. and sent to P.O. Box 206, Ridgecrest, NC 28770. The unit is ideal when radon levels are found to be high in a house, and it is excellent to use while other radon reduction methods are being taken. No-Rad is just 1 foot high, 16 inches in diameter, and weighs 19 pounds. It is a technological breakthrough for radon removal by filtration.

Monitoring the No-Rad

Note that the No-Rad removes only the airborne radon decay products. It does not remove radon gas. As a result, the charcoal canisters, alpha-tracks, and electrets cannot be used to show the radon level reductions. This can be done only with a working level monitor like the Survivor 2, designed to measure the concentrations of airborne radon decay products. No-Rad is more effective when placed as close as possible to the center of the room in which it is located. No-Rad is probably the fastest way to immediately reduce radon levels, since it is simply plugged into an electrical outlet.

THURMOND AIR QUALITY SYSTEMS

Thurmond Air Quality Systems combine matched components into a very high-technology filtration package. They use proven and certified filter media commonly used in hospital operating rooms, "clean" rooms in lab and medical facilities, and in nuclear submarines. In these situations, the cleanest air is mandatory.

When effective air cleaning is used in homes and buildings, lower heating, ventilation, and air conditioning (HVAC) costs are generally realized. Since March 1985, Roger C. Thurmond, the president and founder of the Thurmond Development Corporation, and his wife, Edie, have dedicated their lives to improving indoor air quality. Members of the Roger C. Thurmond family suffered for many years from numerous respiratory and other health problems that proved to be environmentally related.

Thurmond IAQ units start with a special cabinet of great strength designed to accommodate various components, depending on what services the homeowner wants performed. In addition to air filtration, Thurmond units may also incorporate heating and cooling functions. Fans and motors are of the highest quality, and Fig. 6-2 outlines the features of a unit.

IAQ UNIT FEATURES

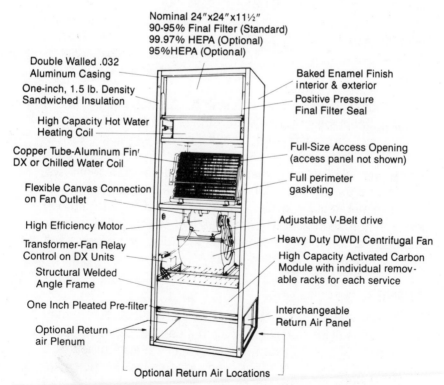

Nominal 24"x24"x11½"
90-95% Final Filter (Standard)
99.97% HEPA (Optional)
95%HEPA (Optional)

Double Walled .032 Aluminum Casing

One-inch, 1.5 lb. Density Sandwiched Insulation

High Capacity Hot Water Heating Coil

Copper Tube-Aluminum Fin' DX or Chilled Water Coil

Flexible Canvas Connection on Fan Outlet

High Efficiency Motor

Transformer-Fan Relay Control on DX Units

Structural Welded Angle Frame

One Inch Pleated Pre-filter

Optional Return air Plenum

Baked Enamel Finish interior & exterior

Positive Pressure Final Filter Seal

Full-Size Access Opening (access panel not shown)

Full perimeter gasketing

Adjustable V-Belt drive

Heavy Duty DWDI Centrifugal Fan

High Capacity Activated Carbon Module with individual removable racks for each service

Interchangeable Return Air Panel

Optional Return Air Locations

Fig. 6-2. Thurmond air quality units remove some radon and many other contaminants in the home.

Thurmond Filters

The heart of the Thurmond IAQ-2000 units is the three-tiered filtration system, which utilizes the best of technology for removal of both particulate and gaseous contaminants. There is a pre-filter, an activated carbon module, and a final filter. The pre-filter is of medium efficiency, a pleated disposable type, made of non-woven cotton and polyester. The activated carbon module consists of racks charged with a minimum of 2,000 cubic inches of activated carbon media. Although activated charcoal's affinity for radon and other contaminants has long been known, there are also different grades of charcoal.

Activated charcoal manufactured from coconut shells is considered to be of the highest quality and used as the standard media in Thurmond units. An impregnated media of specially treated impregnated carbon is used to reduce concentration of trace compounds in the air. A special media of carbon is employed to remove specific targeted contaminants, and a fourth type is multiple media. Seven different racks of these four media make up the carbon module. Each type of carbon media is charged in separate racks to avoid even the slightest contact with one another. Then, a final filter is standard on the Thurmond Systems. This battery of filtration removes practically all contaminants present indoors in the form of gases or vapors.

Activated charcoal is not a cure-all, but it is documented that in a high percentage of cases where there are definite pollutants, positive benefits are achieved. One of the most significant testimonials to the efficiency of activated charcoal in enclosed space atmosphere purification is the space capsule environmental systems. Some 8 pounds of carbon are used in the main adsorbent canister of the Apollo capsules. One cubic foot of activated carbon contains more than 200 million square feet of adsorptive surface. Gas and vapor phase contaminants, including some degree of radon, are deposited on the surface.

OZONE

Just a fraction of a part per million of ozone is detrimental to human health! Ozone is the culprit in the air that causes rust. It can even oxidize paper. Indeed, it was reported recently in a scientific article that ozone in some houses is high enough to cause the deterioration of condoms! This could cause unwanted pregnancies and result in sexually transmitted diseases!

Ozone is a minute fraction of the air and is really another form of oxygen, but it has a tremendously high oxidizing ability. Ozone is produced to one degree or another by electrostatic precipitator-type air cleaners during their normal operation. When they are not operating normally, they may produce large quantities of ozone.

Thurmond Units to Remove Ozone

Thurmond's carbon modules adsorbs ozone and catalytically converts it back to oxygen. Activated carbon has the ability to completely remove ozone from air even when it is present at very low concentrations. If you can smell ozone, it is definitely too high. Ozone causes the tires on cars to crack. Ozone attacks materials and people with more force at higher humidity levels.

Ultibar Technology has a portable building that will stop cars and equipment from rusting, protecting them from ozone and pollutants. The garage-size buildings sell for $995 and offer the ultimate rust protection (FIGS. 6-3 and 6-4). Contact Ultibar at P.O. Box 155, Black Mountain, NC 28711 for information.

The Thurmond IAQ System costs approximately $2,100 for the typical 1500- to 1800-square-foot house, plus installation. Because of the tolerances and clearances needed, the unit should be installed only by professional HVAC people. It is designed to be retrofitted into existing HVAC systems to cure those houses and buildings with "sick building syndrome." Thurmond Systems are ideal for new construction (FIG. 6-5). Write P.O. Box 155, Black Mountain, NC 28711 for information.

Special Thurmond Applications

Although Thurmond air purifications systems are super for houses for relief of allergies and the ills of the chemically sensitive—and an improved quality of life—schools, nurseries, restaurants, smoking areas, and offices are among those that need the most effective IAQ units.

Thurmond units reduce molds, spores, and contaminants in libraries and museums, offering maximum protection to books and artifacts. In dental offices, the Thurmond reduces the concentration of highly toxic mercury gas. Anywhere in industry where people work around toxic substances, the Thurmond should be installed.

For radon removal, the Thurmond is probably as effective as any air purification system on the market. But keep in mind that no

Fig. 6-3. *Ultibar Humidity Control Shelter stops rust by keeping out ozone, preventing auto and equipment corrosion.*

Fig. 6-4. *An interior view of the shelter shows its generous size.*

Fig. 6-5. *Thurmond Air Filtration Systems have superbly engineered components.*

one system or technique is the total answer to reducing radon levels to the Congressional objective of having the indoor level as pure as that of the outside air.

CENTRAL VACUUM SYSTEM

Only a central vacuum cleaner system actually removes dust from rooms. Conventional vacuum cleaners merely stir up dust, and this increases the odds of breathing dust with decaying radon daughters attached to the particles. They knock loose the radon decay products (RDP's), which have been plated out and attached to nonbreathable objects. They get them airborne again!

Electrolux has developed a central vac system that is the most powerful ever made. It has more than twice the suction strength of portable vacs as it removes dust, dirt, and many pollens and pollutants along with them. The central vac is easy to install and is effective in houses up to 7,000 square feet.

Amazingly, the new Electrolux Central Vac System costs less than many portable cleaners! It is available throughout the nation for as little as $699. You may find out where the central vac system is available in your area by calling 1-800-332-3589. In a typical house, this powerful vac may need to be emptied only once or twice yearly, requires no bags, and has a filter that needs washing only once yearly (FIG. 6-6).

When dust is removed, instead of recirculated, the odds of inhaling RDP's are reduced. It also helps remove many allergens, which cause sneezing, coughing, congestion, and all the ailments associated with allergies.

Fig. 6-6. A powerful central vac system like the Electrolux removes dust instead of blowing it back into the room, reducing the chances of radon decay products attaching to particles.

MOLECULAR ADSORBER

An environmentally safe odor and gas adsorber is now available to consumers. Composed of clay mined in New Mexico and Arizona, the molecular sieve grabs odors and scores of pollutants out of the air in homes, offices, and factories. It pulls odors from as far as 300 feet away. Early tests show it is many times more effective in adsorbing radon gas than charcoal.

Molecular Adsorber comes in bottles. Take off the lid, and set the container in a room or area where odors or pollutants are present. A quart may last one year and covers up to 1,000 square feet. The clay is laced with blue granules of sodium cobalt chloride. When they turn blue, the Molecular Adsorber has adsorbed all the odors and fumes it can hold.

People who suffer allergies or upper respiratory ailments may experience a lot of relief by keeping a container in the areas they frequent at home or work. It is being used where offensive odors and pollution are problems, including chemical plants, sewage treatment areas, labs, dark rooms, print shops, gyms, stores, restaurants, and manufacturing facilities. In addition, distilleries, meat plants, laundry rooms, beauty salons, around pets, and even carpet stores are among those using the special adsorber.

Nursing homes, clinics, and other medical facilities are using the molecular sieve. Four-ounce containers are being put in cars and trucks (FIG. 6-7) to keep them odor-free and to adsorb fumes.

Fig. 6-7. Molecular Adsorbers are very effective in cars and trucks, adsorbing odors and fumes.

The adsorbers come in four types and sizes. The 12-ounce plastic jar sells for $16.99, the 32-ounce container for $26.99, the 4-ounce container for $7.99.

Molecular Adsorbers are available from Memories, P.O. Box 206, Ridgecrest, NC 28770. Checks or money orders should be made to Memories, with $2 added for shipping.

In one radon test made by the Department of Energy in Arizona, the Molecular Adsorber adsorbed 272 pCi/l and 13.3 grams of moisture in 48 hours. The same amount of activated charcoal adsorbed 72 pCi/l and 9.9 grams of moisture in the same period.

MONITORS TO CHECK OTHER POLLUTANTS

Carbon monoxide (CO) kills people within minutes when they are exposed to it in heavy doses. CO is much more prevalent than radon. Both are colorless and odorless, but CO kills quickly. CO occurs anywhere fossil fuels are burned. Without sufficient air, CO forms.

Non-fatal levels can cause severe heart and brain damage. CO poisons red blood cells. Low to medium levels may resemble flu, an impending stroke, or severe anxiety. The smaller the person, the quicker the poisoning. CO deaths to children under five have increased 143 percent in the last decade. Some 1,500 people will die this year from accidental CO exposure. Another 10,000 will need medical attention or lose one day of work due to CO. A test kit is available from Memories, P.O. Box 206, Ridgecrest, NC 28770 for $3.95. An electronic gas alarm costs $199 and is available from the North American Wildlife Center, P.O. Box 155, Black Mountain, NC 28711.

The federal government continues to express concern over formaldehyde. All pressed-wood products, particle board, and plywood contain formaldehyde. Newman First Aid distributes a KEM Medical Formaldehyde Monitor. The $949 monitor comes with a prepaid postage mailer. Checks or money orders should be made payable to Newman First Aid and sent to P.O. Box 206, Ridgecrest, NC 28770.

7

Radon and Your Water Supply

Groundwater is another source of indoor radon. As with radon in the soil, the primary risk is inhaling radon that has been released from water into the air. The EPA estimates that water contributes one to seven percent of the radon found inside houses and buildings.

Any process that exposes water to air releases the radon. Most homes are served by public water supplies, which are aerated at the water treatment facilities of cities and towns. Very little radon reaches the majority of homes in America. We release the radon from the water coming into homes by bathing, showering, washing clothes, flushing toilets, and running water from faucets.

Guidelines have been developed to estimate the relationship between radon in water and the resulting level of radon in air. Roughly stated, 10,000 pCi/l of radon in water will lead to 1 pCi/l of radon in indoor air. The concentration of radon in drinking water from public water supplies serving more than 1,000 people is about 240 pCi/l. The average for all public drinking water supplies is about 420 pCi/l. These are negligible unless the radon level inside a house is already at the safe level of 4 pCi/l.

PRIVATE WELLS

Some radon authorities say more than one million wells in America have very dangerous radon levels. Radon-hot wells have been tested with levels as high as two million pCi/l! Houses that receive water from a private or small community well should have the radon level in the water measured.

A qualitative test may be performed by using either grab samples or a continuous radon monitor—like No-Rad—to measure the radon concentrations in a closed bathroom before and after the hot shower runs for 10 to 15 minutes. If the well water contains more than, say, 40,000 pCi/l of radon, the water may be contributing a significant portion of the indoor airborne radon. Local concentrations in the bathroom may be much higher than in the rest of the house. There is no documented health risk from drinking radon in water.

RADON IN DRINKING WATER

Michael Cook, director of the EPA Office of Drinking Water, says radon could be the biggest contaminant in the nation's drinking water. A maximum contaminant level (MCL) has not been set by the EPA, but four figures will probably be considered: 200, 500, 1,000, and 2,000 pCi/l. Some estimates are below 200.

More firms are testing the radon concentrations in water now. One of the biggest firms does tests for $35. (For the test kit and the instructions send a check or money order for $35 to Memories, P.O. Box 206, Ridgecrest, NC 28770.)

"We have done a nationwide survey of radionuclides in drinking water and found there are elevated levels in some drinking water supplies across the country," Cook said.

WATER TO AIR RADON TRANSFER

The amount of radon transferred from water to air depends on the following:

- ‣ The waterborne radon level
- ‣ The amount of water being consumed
- ‣ Water activity use (showering transfers a lot while running water in sinks has a low transfer)
- ‣ The temperatures of the water and the air

"Levels of radon that one can expect to encounter in household well water supplies range from near zero to over one million pCi/l," the EPA Second Edition of *Reducing Radon in Structures*, states. Using the ratio of 10,000 pCi/l (water) to 1.0 pCi/l (air), it takes a waterborne radon level of 40,000 pCi/l to result in the EPA guideline level of 4 pCi/l in the air."

However, the 10,000 to 1 ratio is often not valid, the EPA admits, and water is rarely the sole contributor to a radon level. Yet, the health risk due to radon coming from water is relatively high compared to the maximum contaminant levels (MCL) for many other contaminants found in drinking water!

MEASURING RADON IN WATER

Getting a syringe full of water on the site and then sending it to a lab for counting the radon level by liquid scintillation is one method. Collecting some water in a vial and then having it counted in a lab is another method.

The second method is most popular since the water in the syringe must be transferred and radon de-gassing may occur.

According to the EPA, the homeowner can gather his own sample by following these steps:

1. If there is an aerator on the faucet, remove it.
2. Run the cold water for one minute at a low but steady rate, or until the temperature becomes constant.
3. Slowly fill a 40-ml vial without causing any bubbles to form and allow the vial to spill over for a few seconds.
4. Cap the vial immediately, making sure there are no air bubbles. A Teflon or aluminum seal should be under the lid.
5. Invert the vial. If there are any air bubbles, start over.
6. Record the time, date, and the location where the water sample was obtained.
7. Pack the bottle carefully in some wrapping to ensure it does not get broken.
8. Mail the sample to a lab, preferably overnight delivery to minimize the decay and the counting error. The lab should make its measurement on the day it arrives.

There is a much slower means of measuring the radon content of your water. Terradex makes an alpha-track detector designed to be immersed in a bathroom toilet tank over a period of weeks. The device is available for $31 with a check or money order made payable to Terradex and sent to Memories, P.O. Box 206, Ridgecrest, NC 28770.

AERATION METHODS

Spray, diffused bubble, counter-current-packed-tower, and the horizontally extended tray are the four ways aeration can remove radon from water. Although diffused-bubble aeration has been used for larger water supplies, packed-tower aeration is considered more cost-effective when the water flow is greater than 20 gallons per minute.

There are now some very effective bubble aeration systems on the market that can remove radon to extremely low levels. Although the EPA reports the installed cost of these units for household use ranges from $1,700 to $3,000, there are some units that may be installed for less than $1,700! Well water is sprayed from an inlet into a tank through fine mist spray nozzles. Just a simple spraying of water removes half the radon content immediately. In some cases, the water is resprayed several times when very high radon levels need to be reduced (FIG. 7-1).

SPRAY AERATION

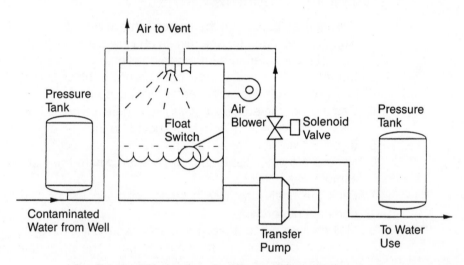

Fig. 7-1. An EPA drawing of a spray aeration system.

Packed columns are filled with about 5 feet of an inert material. Water goes in the top of the column and runs down through the packing, acting like a huge filter. A small blower pushes air up through the packing from the bottom, pushing out the radon gas released from the water through a vent to the outside, as shown in Fig. 7-2. This method is considered highly efficient for low levels—up to 20,000 pCi/l—of radon contamination. For higher contaminations, this system is not considered practical.

PACKED COLUMN

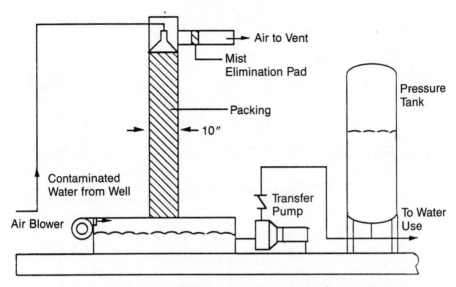

Fig. 7-2. An EPA drawing of a packed column system.

Diffused aeration, as illustrated in Fig. 7-3, utilizes two or more aeration tanks with bubble diffusers in their bottoms. A blower forces air into the water, and as the bubbles rise, the radon volatilizes into the air. Two to six tanks are required for this system. The EPA points out that the holes in the diffuser foul easily due to their smallness.

Shallow aeration uses a tray design. Water is sprayed across a baffled tray inside the aeration tank. Air is blown up through the holes in the tray, frothing the water where the air strips as much as 99.5 percent of the radon, which is vented to the outside of the house (FIG. 7-4). The chief criticism of this method is that it requires a lot of air, some 100 cfm. In some homes, the EPA warns, an outside air source may be needed to prevent basement depressurization caused by the air consumption.

"Recently, several issues have created a situation advantageous to increased use of aeration for radon removal," the EPA states in one of its manuals. "First, it is clear that the EPA intends to set a relatively low maximum-contaminant level for radon, possibly 200 to 2,000 pCi/l. While the MCL will apply directly only to public supplies, it may be interpreted in some instances, such as real

DIFFUSED AERATION

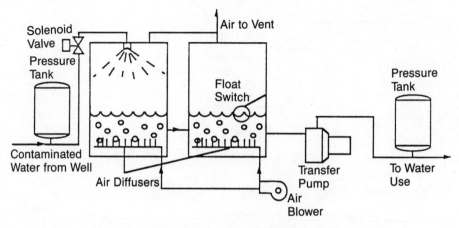

Fig. 7-3. An EPA drawing of diffused aeration.

SHALLOW AERATION

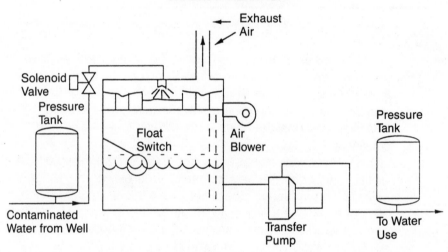

Fig. 7-4. An EPA drawing of shallow aeration.

estate transactions, as a guideline for removal in residences. Second, new developments in point-of-entry and small systems aeration technology have made aeration increasingly effective, comparable to other techniques in removal efficiency."

One of the most impressive aerators is available from Action Manufacturing & Supply, Inc. The Water Guard Aerator is not only effective in radon removal, it is very efficient at taking out hydrogen sulfide, carbon dioxide, carbonic acid, and methane gas.

Action's Water Guard Aerators

The Action Outdoor Aerator consists of a large fiberglass storage tank with an inlet from the well pump to a number of fine mist spray nozzles that mix air with the water as the tank fills. The gases are released by this process from the water. Large screened vents dispense this gas to the atmosphere. Various controls are used to maintain the proper amount of water in the tank. A second pump system draws the water from the tank and repressurizes it to serve the needs of the building. (See FIG. 7-5.)

Fig. 7-5. Water Guard Aerators are exceptionally strong.

For a single-family residence or light commercial installation, the Water Guard Aerator package includes the components listed below. Using an existing well pump and pressure tank, the aerator has an intake capacity of up to 24 gallons per minute (gpm). The jet pump will repressurize the system to 11.7 gpm at 40 psi:

1—A30V Aerator
1—Float Switch

1—R2041 Solenoid Valve, 1 inch
1—C115 Electric cord, 115 V, pump
1—Electric Cord, 115 V, solenoid valve
1—FV100 Foot Valve, 1 inch
1—PG100 Pressure Gauge
1—50RJ ½ HP Jet Pump
1—WX202 Well-X-Trol, 42 gallon

The total cost is less than $1,000, and, because of the simplicity of the system, the installation fee should not be excessive.

The Action Indoor Aerator model includes the drainpipe, draw pipe, manifold, and blower and is approximately the same price as the outdoor model. An accomplished family handyman familiar with basic plumbing may be able to install the system in one day (FIG. 7-6). Water Guard is located at 1226 S.E. 9th Terrace, Cape Coral, Florida 33990. The telephone number is 813-574-3443, and Water Guard technicians will aid well owners in selecting a system to meet their needs.

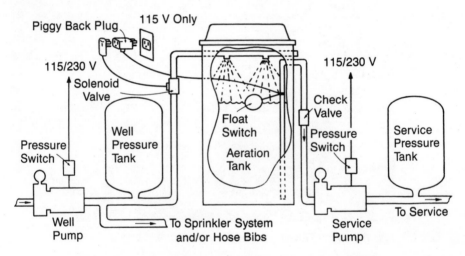

Fig. 7-6. *A layout drawing of a Water Guard Aerator System.*

Water Guard Aerators are built of strong fiberglass with an easily cleaned, sanitary, gel-coated interior. They have many screened vents for better ventilation, and more spray nozzles than most systems for better aeration. They feature extra-strength concave bottoms, strong convex lids, and an exclusive recessed drain. Water Guard Aerators have been installed throughout the nation for years. Water Guard's president, John Guard, is one of the most informed water technology people on aeration and the removal of radon from water.

8

Building a Radon-Resistant House

Radon and real estate are becoming increasingly intertwined. The National Association of Realtors has formally adopted policies to educate homeowners concerning monitoring and mitigating indoor air quality problems.

The Association's Code of Ethics requires full disclosure of any known environmental health hazards by homeowners in listing homes for sale.

BUYERS DEMAND TESTING

Buyers are demanding that homes be tested before they are put up for sale. Radon-testing clauses in real estate contracts are becoming common, and authorities say lending institutions and banks will soon require buildings to meet EPA minimum guidelines before loans are made. Naturally, some real estate people hope radon requirements will not become so rigid and go the way of other indoor air quality problems. Formaldehyde upset the real estate world tremendously for awhile and then subsided; however, indoor air quality authorities say formaldehyde is just as deadly as it ever was.

Building codes and regulatory agencies of government will probably demand that new construction adheres to radon-resistant techniques and materials. Some builders started radon-resistant new construction as soon as radon was identified as a killer.

The three approaches to resolving the radon problem in new construction are preventing radon entry by using barrier methods, reducing the radon entry driving forces, and diverting radon from entering houses by using sub-slab ventilation. We have covered these aspects earlier in detail.

BUILDING SITES

"Although radon in water and radon emissions from building materials account for a minor share of the radon problems in America, the overwhelming majority of residential radon problems occur from radon emanating from the soil," the EPA says of the identification of radon-prone building sites.

The amount of radon gas that enters a house depends on how much radon gas or radon parent compounds are found in the soil under the house, the permeability of the soil, the presence of faults and fissures in the underlying and nearby rock, openings between the house and the soil, and the driving forces that move soil gas containing radon along these pathways into the house.

For a structure to have a radon problem, there must be radium close to it, a way for the gas to move through the soil or rock, a driving force, and openings in the foundation. Being able to determine if a site has a high radon concentration would be ideal. The EPA feels there are no reliable methods for correlating the results of radon soil tests at a building site with the subsequent indoor radon levels in a house built on that site.

Indoor and Soil Radon Concentrations

Having a high radon concentration in the soil on a site may not necessarily mean anything. A Florida radon study matched up more than 2,000 soil radon and indoor radon samples. A total of 77 soil radon readings were greater than 1,000 pCi/l, and the two highest readings were 6587.0 and 6367.2 pCi/l. The indoor readings in the houses sitting on these grounds at different locations were 6.8 and 0.2 pCi/l.

The Florida survey was very enlightening since 95 percent of the houses in the state are slab-on-grade construction. However, one cannot look exclusively at houses built on concrete pads. Below-grade basements and the height and ventilation of crawl spaces must be taken into consideration when considering soil radon concentrations and measurements.

Some sections of the nation do have higher soil radon levels. In Clinton, New Jersey, a homeowner in a subdivision measured the radon level in his house and received a very high radon reading. The New Jersey Department of Environmental Protection surveyed the neighborhood and found that 101 out of 103 houses had radon levels above the EPA action level. More than half the homes had more than 25 times the action level.

Radon is very capricious. One house may have a very deadly radon level, and a house in the same block might be below the 4 pCi/l safe level. In Boyertown, Pennsylvania, the EPA found radon levels in some houses more than 500 times above the safe level. Some of them were next door to homes that had safe readings! Again, this just documents how scattered radon concentrations can be. The presence of elevated-radon-level houses in a neighborhood is just an indication that the probability of having a radon problem has increased.

FOUNDATION WALLS

Below-grade walls may be constructed or poured concrete, masonry, or other materials such as all-weather wood or stone. Poured concrete and masonry block are most commonly used in new construction.

Poured concrete foundation walls are generally constructed to 3-compression strength. Forms are usually held together with metal ties that penetrate the wall. When they corrode, they can allow radon entry. Aside from cracks, utility openings, and penetration at the ties, poured concrete walls can be good radon barriers.

Concrete-block foundation walls may have open cores. They should be closed at the top course. The exterior of masonry walls is coated with a cementitious material. Uncoated block walls are not an effective radon barrier. Building codes in most sections of the nation dictate some type of dampproofing or water control treatment be applied to poured concrete as well as masonry walls. Concrete and cinder blocks are much more porous than poured concrete. Only by waterproofing can radon gas be prevented from moving through them. A bad batch of poured concrete can also be very porous.

Officials of several code organizations have indicated that they strongly favor poured concrete foundation walls instead of masonry walls because of the potential for radon penetration through block walls. Quality poured concrete is superior unless special efforts are made to close blocks to radon penetration.

Some codes require solid tops on masonry walls. Solid, filled, or sealed block tops are desirable for reasons other than radon resistance. Energy conservation, termite proofing, and better distribution of the weight of the structure are additional benefits.

SEALING MASONRY WALLS

Block tops may be sealed at the top course by stuffing paper, wire mesh, or some other material into the cores and then filling the cores with mortar. It is easy to leave gaps with this method, and termite caps or 100-percent solid blocks are more radon resistant. Anchor bolts need not pose a problem. Certainly, it is easier to stick anchor bolts into the open cores of blocks and seal them well with mortar. Laying the next to last course with solid block is a good technique if this is done. Some codes do not allow anchor bolts to be thrust into open blocks, anyway.

Closing block tops enhances soil depressurization, and, as the EPA states, are necessary in many cases for soil depressurization to work. Sealing at block tops and other potential radon entry points may even be sufficient to maintain radon at an acceptable level in houses with weak radon sources.

A house or structure literally sucks at the ground, trying to draw gases out of the soil. Capping off the blocks removes this suction force. Solid block installed as the bottom course of a foundation wall is recommended to keep radon from seeping into block cores around the footing. Weep holes in properly designed and constructed masonry walls are not needed and should be avoided, the EPA says.

There should be no open cores in blocks at access doors to crawl spaces, around ash pit doors, and other openings. No open cores should be allowed to channel radon into a structure.

EPA MANUAL ON NEW HOUSE CONSTRUCTION

The following pages are quoted from the EPA manual *Radon-Resistant Residential New Construction*. The manual was authored by Michael C. Osborne of the Air and Energy Engineering Research Laboratory, Office of Environmental Engineering and Technology Demonstration, and the Office of Research and Development of the U.S. Environmental Protection Agency in Research Triangle Park, NC.

The document is intended for use by residential housing contractors, new-house buyers, state and federal regulatory officials, residential code writers, and other persons as an aid in the design and application of radon-resistant construction in new houses.[1] A complete copy of this manual is available from the EPA Regional Offices and the EPA's Center for Environmental Research Information, Distribution, 26 W. St. Clair Street, Cincinnati, OH 45268.

* * * *

W.S. Fleming and Associates, was employed to develop radon-resistant new construction designs, see that these designs were built in radon-prone areas of New York, and monitor the completed houses for radon. The incremental cost for radon-resistant construction was paid for by NYSERDA and EPA and not by the builder. At least five additional new houses of similar construction and in the vicinity of the radon-resistant houses would serve as control houses to compare with the houses built using the radon-resistant construction techniques.

Currently only 5 of the 15 houses employing radon-resistant construction techniques have been built, although radon-resistant designs have been developed for all 15 houses. Data have not yet been collected on the effectiveness of the radon-resistant designs. The following are the builder's directions for applying radon-resistant techniques in the construction of these new houses with full basements:

Task 1:   Install a continuous airtight plastic film (6-mil polyethylene or equivalent) over the sub-slab aggregate before the slab is poured.

> ‣ Discharge footing drains to daylight or dry well, whenever possible, to avoid introducing radon into an interior sump. If footing drains discharge into an interior sump, provide the sump liner with an airtight lid (that still allows access to service the sump pump).

> ‣ Seal airtight any tears, punctures, slits, or penetrations of the plastic film with builder's tape (3M 8086™ or equivalent). Overlap the edges of any joining of the plastic film by at least 3 in. and seal airtight with builder's tape (3M 8086™ or equivalent).

> ‣ Affix the plastic film to the footing under the expansion board with a troweled-on asphalt coating (Hydrocide 700™ mastic or equivalent) to prevent radon entering the basement from cracks in the footing and from gaps between the footing and plastic film.

---

[1]The foreword to this manual states the following: "The U.S. Environmental Protection Agency (EPA) strives to provide accurate, complete, and useful information. However, neither EPA nor any person contributing to the preparation of this document makes any warranty, expressed or implied, with respect to the usefulness or effectiveness of any information, method, or process disclosed in this material. Nor does EPA assume any liability for the use of, or for damages arising from the use of, any information, methods, or process disclosed in this document."

(Application of Hydrocide 700™ mastic requires washing the surface with water, then removing any standing water.) Position the plastic film between the expansion board and foundation wall and trim the plastic film below the slab level after the expansion board is removed. (If water collects in the foundation wall, as shown by wet lower surfaces, weeping holes may be introduced into the bottom course of concrete blocks to allow water to flow from the foundation wall into the floor/wall gap.)

▶ Provide a strip of asphalt coating on the top of the plastic film, under the slab around the footing, and around any penetrations to prevent radon entering the basement from gaps between the plastic film and the slab.

▶ Install a 2-in. pipe over the plastic film from the floor/wall gap to the inside of the sump liner. If the footing drains discharge into the sump and not to daylight, the 2-in. pipe is capped to prevent radon from the sump entering the floor/wall gap and basement. (If water from the foundation wall collects in the floor/wall gap, the pipe is uncapped and a small water trap is installed allowing the flow of water from the floor/wall gap to the sump, while not allowing the flow of radon from the sump to the floor/wall gap and basement.)

▶ Use the recommended water content in the concrete mix to minimize drying time and reduce shrinkage and cracks in the slab.

▶ Minimize the number of pours. Seal any control joints with polyethylene foam back rod and polyurethane caulk.

▶ Ensure that steel reinforcing mesh, if used, is embedded in (and not under) the slab to help reduce major floor cracks. Reducing major cracks in the slab (as well as footings, block foundation walls, and poured-concrete walls) will reduce the rate of radon entry.

Task 2a: Install a continuous airtight plastic film (6-mil polyethylene or equivalent) around the exterior of the foundation wall from finished grade level to the bottom of the footing.

▶ Affix the plastic film to the foundation wall and footing using a troweled-on asphalt coating (Hydrocide 700™ mastic or equivalent).

▶ Seal airtight plumbing, electrical, or any other penetrations through the plastic film with builder's tape (3M 8086™ or equivalent) and/or troweled-on asphalt coating (Hydrocide 700™ mastic or equivalent).

OR

Task 2b: Install a continuous layer of surface bonding cement (Foundation Coat™ or equivalent) around the exterior foundation wall and footing.

‣ Ensure that plumbing, electrical, or any other penetrations are sealed airtight.

Task 3: Install two courses of termite blocks at the top and bottom of the concrete block foundation wall, one course directly on the footing (cap up).

Task 4a: Provide for venting the footing drains (and sump, if any) to the outside using 4-in. PVC pipes from the footing drains, along the outside of the foundation wall, to finished grade level. Initially, these PVC pipes are to be capped. At least two, and up to four, vents are to be used on opposite sides of the building, venting at least 10 ft. from the nearest window, door, or other opening into the building.

Task 4b: Provide for venting the interior footing drains (and sump, if any) directly through the roof with the largest PVC pipe possible (4-in. diameter minimum). The PVC pipe is to be initially capped at the slab surface, the basement ceiling surface, and on the outside roof surface.

OR

Task 4c: Provide for venting the interior footing drains and/or sump with at least a 4-in. PVC pipe through the rim joist, venting at least 10 ft. from the nearest window, door, or other opening into the building. The PVC pipe is to be initially capped at the slab surface.

VAPOR-SOIL BARRIERS UNDER CONCRETE

Many different types of flexible materials are being used to cover the clean gravel aggregate—No. 57 crushed stone is among the most commonly recommended—before the concrete slab is poured. Until the deadliness of radon was identified, cheap black plastic was often stretched out over the gravel.

In too many instances, the plastic became ripped and badly torn when the concrete was poured over it. Thicker, higher-quality membranes have been standard on more judicious construction. Some of the membranes available are thick and durable and rather heavy. One of the best vapor/soil gas barriers to go under concrete is Soil-Flex from Energy Saver Imports, Inc. (ESI). It is double-aluminum-foil sandwich with a layer of asphalt and a glass scrim web between the metal.

The very lightweight Soil-Flex stops air filtration. It is also excellent as an outside covering for poured concrete or block walls. The woman in Fig. 8-1 easily hefts a roll.

Fig. **8-1.** *Foil-Ray barriers are very light, as this lady demonstrates with a big roll of the membrane.*

INSULATION AND BARRIERS FOR CRAWL SPACES

ESI has a selection of Foil-Ray reflective insulations and barriers. Some are most effective in crawl space applications. They form a tight moisture- and air-infiltration barrier while preventing heat rays from penetrating down beneath the building. For houses with a crawl space or sections of homes with an open bottom: staple, nail, or glue Foil-Ray Insulation or Foil-Ray AF to the bottom of the joists, and tape the seams with aluminum reflective tape or a high-quality sealing tape. Foil-Ray may also be installed in this manner, supporting a conventional mineral wool batt fitted into the cavity. Overlapping enough to ensure there is no air infiltration is important when installing any barrier. Foil-Ray is excellent to seal off the bottom board of manufactured homes, also. Because of its thinness, strength, and flexibility, it applies perfectly to surfaces.

Scissors, a tape measure, sealing tape, and a stapler are all the tools needed to install Foil-Ray in crawl spaces. In addition, the barriers have many other energy-saving uses around the house.

The aluminum surfaces of the membranes may be used on the roof, in the attic, under the concrete slab, and around foundations, as well as in many other places. It is cost-effective and of the highest quality.

ESI is located at 2150 W. 6th Avenue, Unit E, Broomfield, Colorado 80020. The toll-free phone number is 1-800-228-FOIL. Mr. Bowen Hyma, the General Manager, is one of the foremost authorities on vapor barriers and reflective insulations. He can specify the right barrier for radon-resistant construction.

Foil-Ray is also available from Memories, P.O. Box 206, Ridgecrest, NC 28770 at excellent prices. It is a distributor of all Foil-Ray products.

POLYURETHANE SEALANT

The EPA repeatedly emphasizes in its manuals and literature that polyurethane is the finest sealant. It exhibits a lifetime bond with a lifetime flexibility. The basics concerning sealing and polyurethane were covered in Chapter 5.

We recommend Sikaflex Multi-Caulk be used in all sealing applications when building. It is superior polyurethane. Sika Sales Manager Jim Schwartz, one of the world's leading authorities on sealants, caulks, and concrete, emphasizes that tooling is a very important part of the sealing process. The shape of the seal is crucial when using the highest-grade polyurethane like Sikaflex.

The sealant should be shaped so it touches only two surfaces and should be concave in shape. This allows polyurethane to move when necessary, in accordance with natural expansion and contraction. Note the sealant shapes and dimensions that Sika recommends in Fig. 8-2.

Table 8-1 cites the extent of application of 1 gallon of Sikaflex Multi-Caulk per the number of lineal feet, depending on the size of the cracks to be filled.

Sika has just introduced Multi-Seal, an all-weather, self-adhesive aluminum flashing material. It comes in easy-to-use rolls and may be used in many phases of construction. It covers roof gutter joints, seals skylights perfectly, and performs many flashing applications, as well as sealing applications.

AFM RADON GAS ABATEMENT SYSTEM

Several coatings or sealants to cover basement walls and floors have suddenly appeared on the market. Some are very poor, as

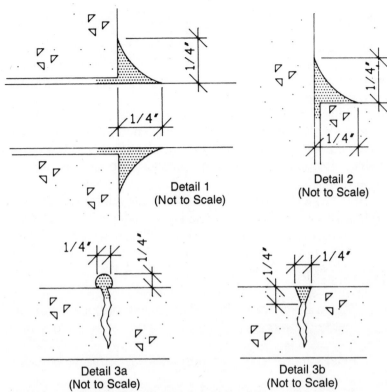

Fig. 8-2. Apply Sikaflex Multi-Caulk polyurethane in these shapes and dimensions when sealing cracks in walls and floors.

Table 8-1. Requirements for Filling Joint Slots

Lineal Feet per Gallon

INCHES

	¹/₄	¹/₂	³/₄	1	1¹/₄	1¹/₂
¹/₄	308.0					
¹/₂	154.0	77.0				
³/₄	102.7	51.3	34.2			
1	77.0	38.5	25.7	19.3		
1¹/₄	61.6	30.8	20.5	15.4	12.3	
1¹/₂	51.3	25.7	17.1	12.8	10.3	8.6
1³/₄	44.0	22.0	14.7	11.0	8.8	7.3
2	38.5	19.3	12.8	9.6	7.7	6.4
2¹/₂	30.8	15.4	10.3	7.7	6.2	5.1
3	25.7	12.8	8.6	6.4	5.1	4.3

radon authorities have told us. Some reportedly contain muriatic acid, which will etch into concrete, and, one expert said, "this is the worst thing in the world you want to do to concrete when you are trying to prevent radon entry." Some of the coating and barrier manufacturers and distributors naturally want to get their products out to the major store chains.

It is basic sales psychology at work. Getting the products out in major discount stores will certainly influence buyers as they are confronted with big sales displays, and people will buy some worthless things thinking they are purchasing good radon-blocking products.

AFM has a Radon Gas Abatement System designed for basements. This same system may be applied to conventional construction and slab floor construction, with a little application adjustment. This is a four-step system, while some on the market utilize only two steps. This system consists of products designated RG1 through RG4:

▸ RG1 is a product that penetrates below the surface approximately $1/16 +$ of an inch and forms a barrier similar to a sheet of glass.
▸ RG2 is a product that encapsulates and forms a membrane over RG1.
▸ RG3 is a black emulsified elastometric membrane system that never loses its elongation, elasticity, or tackiness.
▸ RG4 adds protection but provides a cosmetic effect and a very durable top surface.

All four products in this sealant system may be applied with an ordinary pump-type garden sprayer, brush, or roller. In three tests, done in conjunction with the University of Scranton, only the AFM Radon Gas Abatement System was used. No joints, cracks, or expansion joints were filled, as should ordinarily be done. The purpose was to see just what the sealant system alone would do. When Sika Multi-Caulk is used on all cracks and crevices on the walls and floors in these test houses, the sealing job will be super!

On the first test property, radon was reduced from a high of 27 to 3.2. Property test number two had a high reading of 37.8 reduced to 3.4, and the third test had a property with a reading of 253 reduced to 4.9 when an air system special to the house was not in use. Testing is still underway on this property. (For more information concerning this very special, nontoxic radon gas abatement system, contact Dr. Barbara Williams at P.O. Box 155, Black Mountain, NC 28711. Please enclose $2 for postage.)

The radon gas sealant system costs only 20 cents per square foot, which is cheaper than the very poor gas-blocking treatments mentioned earlier. We highly recommend this system for new and existing homes.

AFM manufactures many environmentally safe products. The line includes paints, cleaners, sealants, coatings, and many others. Shampoos and other personal hygiene products are great for those with allergies.

Appendix
Compiling
the Facts

When EPA personnel inspect homes for radon potential, they have house inspection forms. These surveys help put various types of house designs and construction characteristics in some perspective for studies and reports. They are much more detailed than those used by real estate inspectors because some information is also gathered on the people who live in the house.

The following example of a house inspection form may be utilized for the homeowner's information. Having these facts handy may be very useful in assessing any radon problems that the house may have. See Table A-1.

The next house summary (Table A-2) is the one included in the EPA manual *Reducing Radon in Structures*. This second edition was designed to be used in a three-day training course for "individuals and firms who have an understanding of how radon enters buildings and who are trained in remediation of indoor radon problems." Homeowners and laymen may also benefit from the information.

HOUSE INVESTIGATION SUMMARY

DATE _____   HOUSE ID# _____

ADDRESS _____

_____

INVESTIGATORS _____

Occupant	Age	% Occ	Years Smoked	Smoke Now?	Length of Residency

House Type (X)

☐ Ranch   ☐ Contemporary   ☐ House Age   ☐ Single Family
☐ Split Level   ☐ Bilevel   ☐ # Levels   ☐ Multi Family
☐ Colonial   ☐ Other
☐ Cape, Salt Box

Site Characteristics (X)

☐ Valley   ☐ Level   ☐ Public Sewer   ☐ Public Water
☐ Hilltop   ☐ Slope   ☐ Septic System   ☐ Individual Well

Soil (X)

☐ Clay   ☐ Bedrock Visible
☐ Sand
☐ Loam
☐ Gravelly
☐ Can't tell

Above Grade Construction (% of Total)   *Interior Finish*   *Exterior Finish*

☐ Wood Frame   ☐ Brick   ☐ Attic   ☐ Wood   ☐ Brick
☐ Concrete   ☐ Stone   ☐ Vented   ☐ Sheetrock   ☐ Wood
☐ Concrete Block   ☐ Unvented   ☐ Plaster   ☐ Stucco

Radon Measurement History

Location	Date Tested	Method	Agency	Results
Water Concentration				

Conditions During Site Visit

Outside Temperature _____
Inside Temperature _____
Windspeed _____
Precipitation _____
Snow cover _____

Mechanical Equipment and Stack Effect Bypasses
Exhaust Appliances (number of each)

☐ Range Hood

☐ Cooktop Range

☐ Bath Fan

☐ Dryer to Outside

☐ Large Exhaust Fan

☐ Attic Fan

☐ Heat Recovery Ventilator

Bypasses (S, M, L)

☐ Around Chimney

☐ Balloon Framing

☐ Soffits

☐ Plumbing Case

☐ Recessed Lights

☐ Attic Entry

Heating and Cooling

	Gas	Oil	Electric	Wood	Kerosene	Stack Damper	Air Cleaner	Return Supply	Dedicated Combustion Air (Y,N)
Furnace									
Boiler									
Space									
DHW									
Fireplaces									
Air Conditioning									
Dryer									

☐ Rock Bed Storage

NOTE: Indicate locations of these equipment items on floor plan

Foundation Type (X)

☐ *Crawl Space*

☐ – Heated

☐ – Vented

☐ Connection to Basement?

☐ – Fully Open

☐ – Access Opening

☐ – Access Door

☐ – Isolated or No Basement

☐ Ductwork?

☐ Plumbing?

☐ *Full Basement*

☐ – Heated

☐ – Vented

☐ Door to Exterior

☐ Door to Upper Level

☐ Ductwork?

Average height _____

☐ *Slab-on-Grade*

☐ *Piers*

Soil Exposure (#)

☐ Percent Below Grade

☐ # Walkout Sides

Walls (X)	**Wall Finish (%)**		**Floor (%)**	**Floor Finish (%)**

Walls (X)

☐ Block

☐ Concrete

☐ Stone

☐ Perm. Wood

Wall Finish (%)

☐ None   ☐ Sheetrock

☐ Paint   ☐ Paneling

☐ Stud Wall

Floor (%)

☐ Earth

☐ Slab

☐ Poly

Floor Finish (%)

☐ None

☐ Tile

☐ Carpet

Insulation (Y/N)

☐ Walls

☐ Ceiling

Air Ducts (Y/N)

☐ Supply

☐ – Insulated?

☐ Return

☐ – Insulated?

☐ Sealed?

☐ Supply Under Slab

☐ Return Under Slab

Drainage (Y/N)

☐ Sump Hole

☐ – Sealed

☐ – Interior Pipe

☐ – Exterior Pipe

☐ – Pump

☐ – Water Present

☐ – Drain to Daylight

☐ Interior Footer Drain

☐ Exterior Footer Drain

☐ French Drain

Water (Y/N)

☐ Water Present

☐ Water Damage Present

Potential Entry Points
(S, M, L)   *(Y/N)*

☐ Water Entry	☐ Stairway Framing	☐ Open Block Tops
☐ Sewer Exit	☐ Equipment Supports	☐ Solid Block Tops
☐ Gas Entry	☐ Ash Cleanout	☐ Bathtub
☐ Electric Entry	☐ _____	☐ Shower
☐ Floor Cracks	☐ _____	☐ Toilet
☐ Wall Cracks		☐ Interior Block Walls
☐ Floor-wall Crack		☐ Floor Drains
☐ Pipes thru Wall		☐ _____
☐ Pipes thru Floor		☐ _____
☐ Main House Trap		
☐ Beam Pocket		

Factors Affecting Freezability   **Sub-Slab Material (X)**
(Y/N)

☐ Hot Water Tank	☐ Stone
☐ Cold Water Tank	☐ Sand
☐ Domestic Water Piping	☐ Clay
☐ – Insulated	☐ Bedrock
☐ Heat Distribution Piping	☐ Can't Tell
☐ – Insulated	☐ Poly Sheet
☐ Water Softener	
☐ _____	
☐ _____	

Test Results

Soil Gamma _____   Toilet Bowl Gamma _____

Water "sniffer" test: 30 sec. count _____ x 3.2 = _____ pCi/L

Sub-Slab Vacuum Test			Feet From Suction (Point A)	"Sniffer"			Results under Pressurization (optional) Smoke
Location	Smoke (I/O/N)*	Δ ("WC)		30 sec. average	× 2 = cpm	cpm × 1.6 = pCi/L	

*KEY

I – Smoke Comes into Basement   **N** – No Observable Movement
O – Smoke Leaves Basement

Summary of Sub-Slab Communication (X)

☐ Excellent Over Entire Slab, and Extends to Wall Test Hole

☐ Excellent Over Entire Slab

☐ Fair to Good Over Entire Slab

☐ Good at Perimeter Only

☐ Between Suction Hole F___ and Test Holes _____

☐ Marginal – Fair to Poor

☐ No Observable Communication

Floor Area [   ]   **Volume** – Whole House [   ]

                                              – Basement [   ]

Basement Fan Door Test   **Whole House Fan Door Test**

Airflow Reversal

[   ] CFM   [   ] Shielding Class [   ] ACH at 50 Pascals

[   ] Pascals   [   ] Terrain Class

                         [   ] Wind Speed   [   ] ELA at 4 Pascals

Fan	Basement

Fan	House

RADON SOURCE DIAGNOSIS
BUILDING SURVEY

NAME: _____   HOUSE INSPECTED: (i.d.) _____

ADDRESS: _____   DATE: _____

_____   ARRIVAL TIME: _____

_____   DEPARTURE TIME: _____

PHONE NO.: _____

SURVEY TECHNICIANS: _____

_____

_____

_____

I. Basic Characterization of Building and Substructure

Site

1. Age of house _____

2. Basic building construction:

   Exterior materials _____

   _____

   Interior materials _____

   _____

3. Earth-based building materials in the building - describe:

   _____

   _____

4. Domestic water source:
   a. municipal surface

   b. municipal well

   c. on-site well

   d. other _____

5. Building infiltration or mechanical ventilation rate:

   a. building shell - leaky, moderate, tight

   b. weatherization - caulk, weatherstrip, etc.

   _____

   _____

c. building exposure: (1) heavy forest _____

                     (2) lightly wooded or other nearby buildings _____

_____

                     (3) open terrain, no buildings nearby _____

exhaust fans: (1) whole house attic fans _____

              (2) kitchen fans _____ (4) others _____

              (3) bath fans _____ (5) frequency of use _____

other mechanical ventilation _____

6. Existing radon mitigation measures

Type _____

Where _____

When _____

7. Locale - description: _____

_____

8. Unusual outdoor activities: farm _____

                            construction _____

                            factories _____

                            heavy traffic _____

Substructure

A. Full basement (basement extends beneath entire house)

B. Full crawl space (crawl space extends beneath entire house)

C. Full slab on grade (slab extends beneath entire house)

D. House elevated above ground on piers

E. Combination basement and crawl space (% of each)

F. Combination basement and slab on grade (% of each)

G. Combination crawl space and slab on grade (% of each)

H. Combination crawl space, basement, and slab on grade (% of each)

I. Other - specify

Occupants

A. Number of occupants _____ Number of children _____

B. Number of smokers _____ Type of smoking _____

                                        Frequency _____

Air quality

A. Complaints about the air (stuffiness, odors, respiratory problems, watery eyes, dampness, etc.)

B. Are there any indications of moisture problems, humidity or condensation (water marks, molds, condensation, etc.)? _____

   When _____

Note: Complete floor plan with approximate dimensions and attach.

_____

_____

II. Buildings with Full or Partial Basements

1. Basement use: occupied, recreation, storage, other _____

2. Basement walls constructed of:
   a. hollow block: concrete, cinder
   b. block plenums: filled, unfilled
                     top block filled or solid: yes, no
   c. solid block: concrete cinder
   d. condition of block mortar joints: good, medium, poor
   e. poured concrete
   f. other materials - specify _____
   g. estimate length and width of unplanned cracks: _____
   h. interior wall coatings: paint, sealant, other _____
   i. exterior wall coatings: parget, sealant, insulation (type _____)

3. Basement finish:
   a. completely unfinished basement, walls and floor have not been covered with paneling, carpet, tile, etc.:

   _____

   b. fully finished basement - specify finish materials:

   _____

   c. partially finished basement - specify:

   _____

4. Basement floor materials:
   a. contains unpaved section (i.e., exposed soil) - specify site and location of unpaved area(s):

   _____

   b. poured concrete, gravel layer underneath

c. block, brick, or stone - specify _____

d. other materials - specify _____

e. describe floor cracks and holes through basement floor

_____

f. floor covering - specify _____

5. Basement floor depth below grade - front _____ rear _____ side 1
   _____   side 2 _____

6. Basement access:
   a. door to first floor of house
   b. door to garage
   c. door to outside
   d. other - specify _____

   _____

   _____

7. Door between basement and first floor is:
   a. normally or frequently open
   b. normally closed

8. Condition of door seal between basement and first floor - describe (leaky, tight, etc.):

   _____

9. Basement window(s) - specify:
   a. number of windows: _____
   b. type: _____
   c. condition: _____
   d. total area: _____

10. Basement wall-to-floor joint: _____
    a. estimate total length and average width of joint: _____
    b. indicate if filled or sealed with a gasket of rubber, polystyrene, or other
       materials - specify materials: _____
    c. accessibility - describe: _____

11. Basement floor drain:
    a. standard drain(s) - location: _____
    b. French drain - describe length, width, depth: _____

c. other - specify: _____

d. connects to a weeping (drainage) tile system beneath floor - specify source of information (visual inspection, homeowner comment, building plan, other): _____

e. connects to a sump

f. connects to a sanitary sewer

g. contains a water trap or waterless trap

h. floor drain water trap is full of water:

    (1) at time of inspection

    (2) always

    (3) usually

    (4) infrequently

    (5) insufficient information for answer

    (6) specify source of information: _____

12. Basement sump(s) (other than above) - location: _____

    a. connected to weeping (drainage) tile system beneath basement floor - specify source of information:

    _____

    _____

    b. water trap is present between sump and weeping (drainage) tile system - specific source of information: _____

    _____

    _____

    c. wall of floor of sump contains no bottom, cracks, or other penetrations to soil - describe:

    _____

    _____

    d. joint or other leakage path is present at junction between sump and basement floor - describe:

    _____

    _____

    e. sump contains water:

        (1) at time of inspection

        (2) always

        (3) usually

        (4) infrequently

        (5) insufficient information for answer

        (6) specify source of information: _____

(7) pipe or opening through which water enters sump is occluded by water:

    (a) at time of inspection

    (b) always

    (c) usually

    (d) infrequently

    (e) insufficient information for answer

    (f) specify source of information: _____

  f. Contains functioning sump pump: _____

13. Forced air heating system ductwork: condition of seal - describe:

                             supply air: _____

                             return air: _____

– basement heated: a. intentionally

                    b. incidentally

14. Basement electrical service:

  a. electrical outlets - number _____ (surface or recessed)

  b. breaker/fuse box - location _____

15. Penetrations between basement and first floor:

  a. plumbing: _____

  b. electrical: _____

  c. ductwork: _____

  d. other: _____

16. Bypasses or chases to attic (describe location and size):

  _____

17. Floor material type, accessibility to flooring, etc.: _____

18. Is caulking or sealing of holes and openings between substructure and upper floors possible from:

  a. basement?

  b. living area?

  _____

  _____

III. Buildings with Full or Partial Crawl Spaces

1. Crawl space use: storage, other _____

2. Crawl space walls constructed of:

    a. hollow block: concrete, cinder

    b. block plenums: filled, unfilled

                     top block filled or solid: yes, no

    c. solid block: concrete, cinder

    d. condition of mortar joints: good, medium, poor

    e. poured concrete

    f. other materials - specify: _____

    g. estimate length and width of unplanned cracks: _____

    h. interior wall coatings: paint, sealant, other _____

    i. exterior wall coatings; parget, sealant, insulation (type _____)

3. Crawl space floor materials:

    a. open soil

    b. poured concrete, gravel layer underneath: _____

    c. block, brick, or stone - specify: _____

    d. plastic sheet condition: _____

    e. other materials - specify: _____

    f. describe floor cracks and holes through crawl space floor: _____

    g. floor covering - specify: _____

4. Crawl space floor depth below grade: _____

5. Describe crawl space access: _____

    condition: _____

6. Crawl space vents:

    a. number _____

    b. location _____

    c. cross-sectional area _____

    d. obstruction of vents (soil, plants, snow, intentional) _____

7. Crawl space wall-to-floor joint:

    a. estimate length and width of crack _____

    b. indicate if sealed with gaskets of rubber, polystyrene, other - specify ____

    c. accessibility - describe _____

8. Crawl space contains:

    a. standard drain(s) - location _____

    b. French drain - describe length, width, depth _____

    _____

    _____

    c. sump

    d. connect to: weeping tile system _____

        (1) sanitary sewer

        (2) water trap (trap filled, empty)

9. Forced air heating system ductwork: condition and seal - describe

10. Crawl space heated: a. intentionally

                            b. incidentally

11. Crawl space electrical service:

    a. electrical outlets - number _____

    b. breaker/fuse box - location _____

12. Describe the interface between crawl space, basement, and slab: _____

    _____

13. Penetrations between crawl space and first floor:

    a. plumbing: _____

    b. electrical: _____

    c. ductwork: _____

    d. other: _____

14. Number and locations of bypasses or chases to attic: _____

15. Caulking feasible from: a. basement

                            b. living room

IV. Buildings with Full or Partial Slab Floors

1. Slab use: occupied, recreation, storage, other: _____

2. Slab room(s) finish:

    a. completely unfinished, walls and floor have not been covered with

       paneling, carpet, tile, etc.

    _____

b. fully finished - specify finish materials _____

c. partially finished - specify _____

3. Slab floor materials:

    a. poured concrete

    b. block, brick, or stone - specify _____

    c. other materials - specify _____

    _____

    _____

    d. fill materials under slab: sand, gravel, packed soil, unknown

       source of information _____

    _____

    e. describe floor cracks and holes through slab floor _____

    f. floor covering - specify _____

4. Elevation of slab relative to surrounding soil (e.g., on grade, 6 in. above

    grade): _____

    – is slab perimeter insulated or covered? yes, no

5. Slab area access to remainder of house - describe: _____

    – normally: open, closed

6. Slab wall-to-floor joint (describe accessibility):

    a. estimate length and width of crack _____

    b. indicate if sealed with gasket of rubber, polystyrene, other - specify ____

    c. accessibility - describe _____

7. Slab drainage:

    a. floor drain - describe _____

    b. drain tile system beneath slab or around perimeter - describe _____

    c. source of information _____

8. Forced air heating system ductwork:

    a. above slab condition and seal - describe _____

    b. below slab: _____

        (1) length and location _____

        (2) materials _____

9. Slab area electrical service:

    a. electrical outlets - number _____

    b. breaker/fuse box - location _____

10. Describe the interface between slab, basement, and crawl space:

    _____

11. Penetrations between slab area and occupied zones:

    a. plumbing _____

    b. electrical _____

    c. ductwork _____

    d. other _____

12. Bypasses or chases to attic: _____

    _____

    _____

V. Substructure Service Holes and Penetrations
(Note on floor plan)

Complete table to describe all service penetrations (i.e., pipes or conduit for water, gas, electricity, or sewer) through subfloors and walls. Indicate on floor plan.

Description of service, size, location, accessibility	Size of crack or gap around service and type and condition of seal
Example: water, 3/4-in. copper pipe, through floor, accessible.	*Example:* Approx. 1/8-in. gap around circumference of pipe with sealing polystyrene gasket.

Index

A

Action Outdoor water supplies aerator, 104-106
aeration of water supplies, 100-106
    Action Outdoor, 104-106
AFM radon gas abatement system, 115-118
AirXchange Inc., 81
Apollo HydroHeat, 67-70
attic ventilation, 73-75

B

baseboard depressurization, 51-57
basement pressurization, 57-62
  Kanalflakt Fans, 60-62
block walls, depressurization of, 48-51
Blue Angel sump pumps, 67
building radon-resistant homes, 107-118
    AFM radon gas abatement system, 115-118
  building site testing, 108
  EPA manual for, 110-113
  foundations, 109
  insulation and barriers for crawl spaces, 114
  polyurethane sealant, 115
  sealing masonry walls, 110
  testing, 107-108
  vapor-soil barriers under concrete, 113

C

carbon monoxide, viii, 96
caulk, 116
central vacuum systems, 94
charcoal analysis, 14-16
Congressional action, ix
continuous radon monitors, 18
Cook, Michael, 99
crawl spaces, insulation and barriers for, 114

D

depressurization removal, 36-37
    baseboard, 51-57
    block walls, 48-51
    plastic film, 57
dilution, 72-85
direct vent fireplace, 70-71
direct vent heating systems, 67-68
drain sealing, 63-65
    Dranjers for, 64-65
drain tile soil ventilation, 46-48
Dranjers, 64-65

E

electret-passive environmental monitor, 8-9
electrets, 17

entry points, 1-2, 23-33
  air pressure differences, 24
  contaminated materials, 28-29
  foundations, 29-33
  man-built pathways, 27-28
  winter, 25
EPA action level, 13-14

F

filtration, 86-97
fireplaces, direct vent, 70-71
follow-up measurement, 11-13
foundations, 29-33, 109

G

Geiger counters, 20
government warnings, 3-4

H

half-life, 2
Harley, Naomi, 6
heat recovery ventilation, 78-85
  AirXchange Inc., 81
  forced-air system interface for, 80
heating systems, direct vent, 67-68
house investigation summary, 120-125

I

indoor concentrations, 33-34
insulation, 114
ionization, radon and, 87

K

Kanalflakt Fans, 60-62

L

lead, ix
Leslie-Locke rotary turbines, 77
lung cancer, 1-7

M

measurement, 2
  follow-up, 11-13
  in water, 100
  passive, 9-10
  screening, 10-11
molecular adsorbers, 94-96

N

naturally powered ventilation, 72-73
nitrogen dioxide, ix
No-Rad, 86, 88

O

ozone, viii, 91

P

particulates, viii
passive measurement, 9-10
passive ventilation, 73
plastic film, depressurization under, 57
polyurethane sealant, 115
power ventilators, 75-76
pressurization, basement, 57-62

private wells, 98-99

R

radon
  compiling facts on, 119-126
  dilution to remove, 72-85
  entry points, 1-2, 23-33
  filtration to remove, 86-97
  government warnings, 3-4
  half-life, 2
  indoor concentrations, 33-34
  ionization and, 87
  lung cancer and, 1-7
  measurement, 2
  redirection to remove, 36-71
  risk evaluation chart, 7
  smokers vs. nonsmokers, 4
  testing, 8-22
  toxic components of, 2-3
  ventilation and, 34-35
  water supply and, 98-106
  water to air transfer, 99
radon source diagnosis building survey, 126-136
Radon-One Charcoal monitors, 16-17
Radon-Resistant Residential New Construction, 110-113
redirection, 39-71
removal, dilution, 72-85
  attic ventilation, 73-75
  heat recovery ventilators, 78-85
  Leslie-Locke rotary turbines, 77
  natural ventilation, 72-73
  passive ventilation, 73
  power ventilators, 75-76
  soffit and undereave vents, 77
  warm-air furnace ventilation, 77
removal, filtration, 86-97
  central vacuum systems, 94
  ionization and radon, 87
  molecular adsorbers, 94-96
  monitors to check other pollutants, 96-97
  No-Rad ionizer, 88
  ozone, 91
  Thurmond Air Quality Systems, 89-91
removal, redirection, 36-71
  basement pressurization, 57-62
  depressurization, 36-27
  depressurization, baseboard, 51-57
  depressurization, block walls, 48-51
  depressurization, plastic film, 57
  direct vent fireplace, 70-71
  direct vent heating systems, 67-68
  drain tile soil ventilation, 46-48

sealing standard drains, 63-65
sealing, 44-45
sub-slab soil ventilation, 38-44
sump pumps, 65-67
rip-offs, x
risk evaluation chart, 7

S

Schwartz, Jim, 115
screening measurement, 10-11
sealing, 44-45
   caulking, 116
   joint slots, 116
   masonry walls, 110
   polyurethane sealant, 115
   standard drains, 63-65
smoking, radon and, 4-6
soffit and undereave vents, 77
sub-slab soil ventilation, 38-44
sulfur dioxide, viii
sump pumps, 65-67
Survivor 2 continous monitor, 19

T

tamper-resistant monitor cages, 21-22
Terradex Alpha Track Detector Kits, 17
testing, 8-22
   before buying home, 107
   building sites, 108
   charcoal analysis, 14-16
   continuous monitors, 18
   electret-passive environmental monitor, 8-9
   electrets, 17
   EPA action level, 13-14
   follow-up measurement, 11-13

Geiger counter, 20
passive measurement, 9-10
Radon-One Charcoal monitors, 16-17
screening measurement, 10-11
Survivor 2, 19
tamper-resistant monitor cages, 21-22
Terradex Alpha Track Detector Kits, 17
Thurmond Air Quality Systems, 89-91
   ozone removal, 91

V

vapor-soil barriers, 113, 114
ventilation, 34-35
   attic, 73-75
   drain tile soil, 46-48
   heat recovery, 78-85
   naturally powered, 72-73
   passive, 73
   power, 75-76
   soffit and undereave vents, 77
   sub-slab soil, 38-44
   warm-air furnace, 77

W

warm-air furnace, ventilation through, 77
water supplies, 98-106
   aeration methods, 100-106
   drinking water radon content, 99
   measurement of, 100
   private wells, 98-99
   water to air radon transfer, 99
wells, 98-99